Springer Geography

For further volumes:
http://www.springer.com/series/10180

Ashraf M. Dewan

Floods in a Megacity: Geospatial Techniques in Assessing Hazards, Risk and Vulnerability

Springer

Ashraf M. Dewan
Department of Spatial Sciences
Curtin University
Bentley, Australia

ISBN 978-94-007-5874-2 ISBN 978-94-007-5875-9 (eBook)
DOI 10.1007/978-94-007-5875-9
Springer Dordrecht Heidelberg New York London

Library of Congress Control Number: 2013930017

Printed on acid-free paper

Springer is part of Springer Science+Business Media (www.springer.com)

Dedicated to my Parents, with love

Foreword

According to the UNDP, in their publication "A Global Report: Reducing Disaster Risk: A Challenge for Development," Bangladesh is the country most vulnerable to tropical cyclones and the sixth most vulnerable to floods of all types. Dhaka megacity, the capital of Bangladesh, is urbanizing rapidly as a result of a number of pressures both of population growth and internal migration.

Recent climatic trends have also added pressure to the situation with increased rainfall and cyclonic activity in the area. Unfortunately this is not the only megacity, or large conurbation in the world, to occupy a vulnerable position. Events in the Gulf of Mexico such as the effects of Hurricane Katrina have shown that vulnerability is not limited to the Third World. It is, however, in the Third World that the effects of such disasters are most devastating. This is due to a combination of low response capacity, poor infrastructure, and rampant poverty.

Since the launch of the first Landsat satellite in 1972, mankind has had an unprecedented capability to map and monitor events such as flooding, and over the past several decades, this has improved with new sensors with higher spatial and temporal resolutions, as well as the availability of spaceborne radar images. Data from these satellite sources, when used in conjunction with other geospatial information such as digital elevation models, biophysical and demographic data, provides us with a powerful suite of tools for disaster monitoring, response, and mitigation.

This book provides the reader with a case study, showing how geospatial tools can be used not only to map and monitor flood events but also in the development of mitigation and response strategies.

Curtin University
Perth, Western Australia

Robert J. Corner

Preface

I experienced the 1988 flood in Dhaka when I was a student in high school. My family members were unable to leave home for more than 3 weeks due to the widespread flooding. I have witnessed how a flood can paralyze the regular functioning of an urban system and make people distressed. We did not face problem with food and shelter because we resided on the second floor of a building in old Dhaka; however, many people suffered immensely from lack of work, food, and shelter. In 1998, I had just finished my postgraduate research in geography when Dhaka again experienced an unprecedented flood that exceeded the previous record of 1988 in terms of scale and intensity. Since then, I have been interested in working with water resources and fluvial systems, particularly floods.

My desire came true when I obtained an opportunity to research floods in higher degree by research. When I started my research, I discovered that my academic supervisor was inclined to examine the physical aspects of flood hazards. However, being a geographer, I was enchanted by the idea of combining both the social and physical aspects of flood hazards. I have been inspired by the works of Professor Susan L. Cutter, Professor Kenneth Hewitt, Professor James K. Mitchell, Professor Mark Pelling, Dr. Ben Wisner, and others who have researched the explicit interpretation of nature–society interactions in assessing natural hazards. These works have enlightened me and given me a solid foundation to incorporate the human dimension into my work. This book is the culmination of my journey that formally started in 2002.

Floods are the most expensive and debilitating natural hazards in the world, particularly in Bangladesh. To save lives and property, flood management has been heavily dominated by engineering solutions. Although a degree of success has been reported, flood still remains an inordinate threat to people, and it is blamed for environmental degradation across the country. As Bangladesh is one of the poorest countries in the world, it is difficult for the country to bear the massive expenses related to flood-control structures. Similar to the other parts of the world, a shift from classic flood protection towards flood risk management was strongly felt after the severe flood in 1998 in Bangladesh, and mapping hazards and vulnerability assessment gained considerable interest among academics and professionals.

Due to increasing concentration of population and infrastructures in urban areas, flood has become a significant threat to many parts of the world, particularly cities in developing countries that accommodate more than ten million inhabitants. This book is concerned with floods in megacities with a particular focus on the megacity of Dhaka, which is the capital of Bangladesh. As Dhaka is expanding rapidly, flood-related damage is increasing at the same time, ruthlessly affecting urban inhabitants and the economy. Although a large number of resources have been invested to ameliorate flood loss, flood damage continues to mount in the city. The situation is likely to worsen, as precipitation is projected to increase in the coming days.

The flood problem is further intimidating by the absence of historical patterns of inundation and damage statistics. In such a position, it is difficult to formulate the appropriate countermeasures to combat the growing threat of flood. To develop baseline information that is crucial for saving lives and property from natural hazards such as floods, this book makes use of geospatial techniques. Although data availability is a serious problem in Bangladesh, I have sourced information from census geography, spatial database, and remote sensing. Creating spatial information of hazards, risk, and vulnerability in relation to floods not only provides baseline information on floods at the megacity scale, but it is also useful in identifying the causes of vulnerability, the spatial distribution of vulnerable communities, and risk areas at the community level.

I owe gratitude to many people who have helped me to produce this book. First, I wish to express my sincere gratitude to Professor Makoto Nishigaki, Professor Takashi Kumamoto, Professor Yasushi Yamaguchi, Dr. Md. Monirul Islam, Dr. Joseph Awange, Dr. Robert Corner, Dr. Md. Humayun Kabir, Dr. Yohei Kawamura, Dr. Maminul H. Sarker, Dr. Zia Rahman, Dr. Sirajur R. Khan, and other colleagues who supported me during my journey. A special thank you goes to SPARRSO, GISTDA, GEOEYE Foundation, CEGIS, and RAJUK for their data and technical assistance. I gratefully acknowledge partial funding from the International Foundation for Science (IFS) in Sweden.

I thank the following for permission to reproduce figures: Academic Press, Sage Publications, Taylor and Francis, John Wiley & Sons, Pearson, Dr. Ben Wisner, the United Nations, the World Bank, and Stanford University. Acknowledgements are also due to people at Springer, especially Dr. Elodie J. Tronche and Corina van der Giessen, for their constant support.

I am obliged to my student Akiko Masuya of Curtin University in Western Australia. Without her support, this book may not have been completed. Thanks are also due to my fellow friends who have been supportive during field visits and other circumstances, particularly B.M. Rabiul Islam, Md. Azizul Islam, Md. Ruhul Amin Shiper, Habibul Bashar, Md. Nazmul Hassan, Md. Zahangir Alam, Rafiq Uddin Ahmed, Md. Farhad H. Sikder, Abdullah-Al-Masud, Md. Zainul Abdein, Md. Basir Ullah, Md. Mohiuddin (Late), Md. Monirul Islam Golder, Asaduzzaman Pramanik, and Maki Ozawa. I also wish to extend my gratitude to the interviewees and informants who generously helped me during my field studies. I also wish to thank anonymous reviewers for their constructive comments and suggestions.

I thank my parents, my sisters and brothers, and my family, whose endless support enabled me to be where I am today. My heartiest thanks to my daughters Mahmuda Sayma and Samika Neha and my son Arham Rayyan for their encouragement, great sense of humor, and enormous patience. I could not have completed this book without their constant encouragement and support. Last but not least, no acknowledgement of this nature would be complete without thanking my wife, Kamrun Nahar, for her constant and endearing support in abundance.

Perth, Western Australia
September 2012

Ashraf M. Dewan

Contents

1 Introduction ... 1
1.1 The Context ... 1
1.2 Floods in Bangladesh ... 4
1.3 Vulnerability of Megacities to Natural Hazards ... 10
1.3.1 Urban Flooding: A Global Phenomenon ... 15
1.3.2 Causes and Types of Urban Flooding ... 17
1.4 Outline of the Book ... 20
References ... 21

2 Hazards, Risk, and Vulnerability ... 35
2.1 Introduction ... 36
2.2 Concept of Vulnerability, Hazard, and Risk ... 36
2.3 Models of Vulnerability and Disaster Risk Reduction ... 40
2.3.1 Pressure and Release (PAR) Model and Access Model ... 40
2.3.2 Hazards-of-Place (HOP) Model ... 41
2.3.3 Regions of Risk Model ... 43
2.3.4 Disaster Risk Reduction Model ... 43
2.3.5 Integrated Assessment of Multi-hazards Model ... 44
2.3.6 The UNISDR Framework for Disaster Risk Reduction ... 44
2.4 Conceptual Framework ... 45
2.5 Vulnerability and Risk Assessment with Geospatial Techniques ... 47
2.5.1 Flood Mapping and Monitoring ... 50
2.5.2 Flood Damage Assessment ... 56
2.5.3 Flood Hazards, Vulnerability, and Risk Assessment ... 57
2.6 Summary ... 62
References ... 62

3 Vulnerability of a Megacity to Flood: A Case Study of Dhaka 75
3.1 Introduction 75
3.2 Physical Environment 76
3.3 Built and Socioeconomic Environments 76
3.4 Floods in Dhaka: History and Causes of Vulnerability 79
3.4.1 Low Elevation 82
3.4.2 Unplanned Urbanization 82
3.4.3 Dilapidated Drainage Systems 84
3.4.4 Poor Performance of Flood Control Works 85
3.4.5 High Population Density and Extreme Poverty 87
3.4.6 Flood Preparedness 90
3.4.7 Sociopolitical Structure 91
3.4.8 Governance Quality 92
3.4.9 Impact of Climate Change 93
3.4.10 Absence of Risk Communication Tools 93
3.5 Summary 94
References 94

4 Spatial and Temporal Distribution of Floods 103
4.1 Introduction 103
4.2 Analytical Techniques 104
4.2.1 Data 104
4.2.2 Satellite Data Preprocessing 106
4.3 Assessing the Accuracy of Flood Maps 114
4.4 Spatial and Temporal Distribution of Floods 114
4.5 Accuracy of Flood Maps 119
4.6 Flood Damage Estimation 119
4.7 Discussions 121
4.8 Summary 124
References 124

5 Modeling Flood Hazards 129
5.1 Introduction 129
5.2 Analysis of Flood Frequency 130
5.3 Analysis of Floodwater Depth 130
5.4 Development of the Flood Hazard Map 133
5.5 Summary 136
References 137

6 Vulnerability and Risk Assessment 139
6.1 Introduction 139
6.2 Analytical Techniques 140
6.2.1 Data 140
6.3 Vulnerability Assessment 141
6.3.1 Analysis of Physical Vulnerability 144
6.3.2 Analysis of Social Vulnerability 149
6.3.3 Analysis of Coping Capacity 156

6.4 Derivation of Composite Vulnerability Index (CVI) 162
6.5 Flood Risk Assessment 162
6.6 Results and Discussion 163
6.6.1 Physical Vulnerability 163
6.6.2 Social Vulnerability 163
6.6.3 Coping Capacity 165
6.6.4 Composite Vulnerability 167
6.6.5 Flood Risk Mapping 167
6.7 Summary 172
References 172

7 Conclusions and Recommendations 179
7.1 Introduction 179
7.2 Major Findings 180
7.2.1 Chapter 1 180
7.2.2 Chapter 2 181
7.2.3 Chapter 3 182
7.2.4 Chapter 4 183
7.2.5 Chapter 5 183
7.2.6 Chapter 6 184
7.3 Limitations of the Work 185
7.4 Application of This Research and Further Research Demand 185
Reference 187

Appendices 189

Index 193

Acronyms and Definitions

Acronyms

AR	Assessment report
AVHRR	Advanced Very High Resolution Radiometer
BBS	Bangladesh Bureau of Statistics
BDT	Bangladesh Taka (*name of Bangladeshi currency*)
BWDB	Bangladesh Water Development Board
CCI	Coping capacity index
CEGIS	Centre for Environment and Geographic Information System
CRED	Centre for Research on the Epidemiology of Disasters
CVI	Composite vulnerability index
DAP	Detailed area plan
dB	Decibel
DCC	Dhaka City Corporation
DEM	Digital elevation model
DL	Danger level
DMA	Dhaka Metropolitan Area
DMB	Disaster Management Bureau
DN	Digital number
EM-DAT	Emergency Events Database
ENSO	El Niño–Southern Oscillation
ERTS-1	Earth Resources Technology Satellite-1
ETM+	Enhanced Thematic Mapper Plus
ERS	European Remote-Sensing Satellite
FAO	Food and Agriculture Organization
FFWC	Flood Forecasting and Warning Centre
GBM	Ganges–Brahmaputra–Meghna
GCP	Ground control points
GDP	Gross domestic product
GIS	Geographic information system

GISTDA	Geo-Informatics and Space Technology Development Agency
HWL	Highest water level
IDNDR	International Decade for Natural Disaster Reduction
IFRC/RC	International Federation of the Red Cross and Red Crescent
IPCC	Intergovernmental Panel on Climate Change
IRS	Indian Remote Sensing
IRSO	Indian Remote Sensing Organization
ISDR	International Strategy for Disaster Reduction
ISODATA	Iterative self-organizing data
LDC	Less-developed country
LIDAR	Light detection and ranging
MESSR	Multispectral Electronic Self-Scanning Radiometer
MLC	Maximum likelihood classifier
MODIS	Moderate Resolution Imaging Spectroradiometer
MSS	Multispectral scanner
PCA	Principle component analysis
PVI	Physical vulnerability index
RADAR	Radio detection and ranging
RAJUK	Capital Development Authority
RMSE	Root-mean-square error
RS	Remote sensing
SAR	Synthetic aperture radar
SOB	Survey of Bangladesh
SPARRSO	Space Research and Remote Sensing Organization
SPOT	Système Probatoire d'Observation de la Terre
SVI	Social vulnerability index
TM	Thematic Mapper
UN	United Nations
UNCHS	United Nations Centre for Human Settlements
UNDP	United Nations Development Programme
UNDRCO	United Nations Disaster Relief Coordinator
UNEP	United Nations Environment Programme
UN-HABITAT	United Nations Human Settlements Programme
WB	World Bank
WFP	World Food Programme
WiFS	Wide field sensors
WL	Water level

Definitions

Katcha houses are generally made from mud, straw and bamboo with corrugated tin rooves.

Mahalla is the smallest urban geographic unit.

Mauza is the smallest rural geographic revenue unit having Jurisdiction List Number (JLN).

Other Urban Area (OUA) includes *thana/upazila* headquarters and also development centres having urban characteristics.

Paurashava is a municipality incorporated and administered by local government under Paurashava Ordinance, 1977.

Pucca houses are constructed with solid walls rooftops.

Semi-pucca houses are constructed with brick-wall and corrugated tin roof.

Thana is a police jurisdiction area and comprising about 100 villages or several city wards. A district consists of several *thanas*.

Union is a rural administrative geographic unit comprising of one or more *mauzas* and villages and governed by Union Parishad Institution.

Upazila is the second lowest tier of regional administration in Bangladesh.

Village is the lowest rural geographic unit either equivalent to a *mauza* or part of a *mauza*.

Ward is an urban administrative geographic unit comprising of one or more *mahallas* and governed by Ward Council.

List of Figures

Fig. 1.1 Number of reported flood events between 1950 and 2011 2
Fig. 1.2 Economic losses from floods, 1950–2011 3
Fig. 1.3 Spatial distribution of flooding in South Asia, 1950–2011 5
Fig. 1.4 Growth rates of urban agglomerations, 1970–2011 11
Fig. 1.5 Urban agglomerations by size class and potential risk of flooding, 2011 17
Fig. 1.6 Effects of urbanization on floods 18

Fig. 2.1 Risk triangle 39
Fig. 2.2 Pressure and release model 40
Fig. 2.3 Hazards-of-place model 42
Fig. 2.4 Conceptual framework for disaster risk reduction 43
Fig. 2.5 Schematic representation of the conceptual framework 45

Fig. 3.1 Location of study area 80

Fig. 4.1 Radarsat ScanSAR raw and filtered images (**a**) Raw image, (**b**) Gamma–MAP filtered image 107
Fig. 4.2 Location of land–water threshold profiles 109
Fig. 4.3 Land–water profile for determining the threshold value 110
Fig. 4.4 Flood depth maps for 1988, 1998, 2004, and 2007 111
Fig. 4.5 Percentage of net inundated areas in Dhaka, 1988–2009 116
Fig. 4.6a Spatial patterns of flooding in Dhaka megacity, 1988–2003 117
Fig. 4.6b Spatial patterns of flooding in Dhaka megacity, 2004–2009 118
Fig. 4.7 Depth–damage curves for housing types (**a**) 2004 flood, (**b**) 2007 flood 120
Fig. 4.8 Depth–damage curves for major roads during 1998 and 2007 floods 120
Fig. 4.9 Estimated financial damage maps based on housing types in the 2004 and 2007 floods, by community 121

Fig. 4.10 Backscattering mechanisms in urban areas: (**a**) Single bounce scattering; (**b**) Double bounce scattering; (**c**) Triple bounce scattering 122

Fig. 5.1 Concept of flood-affected frequency 131
Fig. 5.2 (**a**) Flood-affected frequency and (**b**) floodwater depth 132
Fig. 5.3 Calculation of flood hazard 134
Fig. 5.4 Flood hazard map derived from flood-affected frequency and floodwater depth simultaneously 135

Fig. 6.1 Flowchart showing overall methods of the study 142
Fig. 6.2 Biophysical variables 145
Fig. 6.3 Distance categories according to different flood hazard classes, by area 146
Fig. 6.4 Elevation in each hazard zone, by area 147
Fig. 6.5 Demographic indicators 150
Fig. 6.6 Distribution of socioeconomic indicators 151
Fig. 6.7 Distribution of lifeline and infrastructure indicators 152
Fig. 6.8 Decision hierarchy model for social vulnerability assessment 153
Fig. 6.9 Distribution of coping capacity indicators 158
Fig. 6.10 Decision hierarchy model for coping capacity analysis 159
Fig. 6.11 Estimation of mean housing price 160
Fig. 6.12 Spatial patterns of physical and social vulnerability, and coping capacity, by community 164
Fig. 6.13 Spatial patterns of the CVI, by community 168
Fig. 6.14 Flood risk map, by community 170

List of Tables

Table 1.1 Number of reported disasters and humans affected, 1985–2009 2
Table 1.2 Deaths from floods, by continent, 1950–2011 4
Table 1.3 Effects of floods on humans and the economy in South Asia, by country, 1950–2011 5
Table 1.4 Effects of floods on Bangladesh during *abnormal* events 7
Table 1.5 Temporal distribution of population in top 10 megacities since 1990 (millions) 12
Table 1.6 Top 10 largest city populations in 2011, at 8th–10th risk deciles by hazard type 16

Table 3.1 Demographic characteristics of Dhaka megacity 77
Table 3.2 Slum clusters and slum population in Dhaka, 1974–2005 90

Table 4.1 Attributes of satellite data 105
Table 4.2 Flood damage function for housing types and major roads 113
Table 4.3 Inundation area percentages in Dhaka, 1988–2009 115
Table 4.4 Overall accuracy and kappa coefficient of satellite-derived flood maps, 1988–2009 119

Table 5.1 Classification of flood-affected frequency 133
Table 5.2 Area occupied by each category of floodwater depth 133
Table 5.3 Ranking matrix for two thematic hydraulic parameters 134
Table 5.4 Exposure of the population to flood hazards 136
Table 5.5 Percentage of human settlements in different flood hazard zones. 136

Table 6.1 Semantic scale of the AHP method 143
Table 6.2 Example of a pairwise comparison matrix 144
Table 6.3 Random average consistency indexes for various *n* 144
Table 6.4 Biophysical vulnerability decision hierarchy model 147
Table 6.5 Decision hierarchy model for socioeconomic and infrastructure vulnerability 154
Table 6.6 Decision hierarchy model for coping capacity criteria 161
Table 6.7 Distribution of variables in relation to physical vulnerability 165

Table 6.8 Distribution of variables regarding the social vulnerability zone 166
Table 6.9 Distribution of variables relating to CVI 169
Table 6.10 Distribution of variables in different flood risk zones 171

List of Plates

Plate 3.1	Floodplain development along the embanked part of Dhaka	83
Plate 3.2	Littering garbage and waste to the surface water bodies	85
Plate 3.3	Effort of repairing a flood control structure during 1998 flood in eastern Dhaka	86
Plates 3.4–3.6	Sufferings of urban dwellers due to pluvial flooding in 2012	88
Plate 3.7	Floodwater engulfed Bangladesh Secretariat at the center of Dhaka	89
Plate 3.8	Informal settlements in Dhaka	89
Plate 3.9	Environmental quality of a slum	90
Plate 3.10	Vulnerability of marginal group to flooding	91

Chapter 1
Introduction

Abstract This chapter presents trends in flood occurrences and associated damage since the 1950s in the context of the world, South Asia, and Bangladesh. The types of urban flooding and their causes are discussed to shed light on the increasing vulnerability of megacities to floods. The analyses reveal that floods and associated losses around the world have increased with time. However, while flood-related fatalities have decreased substantially, economic losses have increased, disproportionately affecting developing countries. Among the South Asian countries, India had the highest occurrence of floods, followed by Bangladesh. The vulnerability of megacities to floods, particularly in developing nations, is exacerbated due to rapid urban expansion, increasing concentration of population and property, rampant poverty, physical location, and poor-quality housing.

1.1 The Context

Flood is the most expensive and devastating natural hazard (Wilby and Keenan 2012; Sanyal and Lu 2004), and it continues to be a concern in many parts of the world (Jha et al. 2012; Kundzewich et al. 2010; Chang and Franczyk 2008). For instance, the Intergovernmental Panel on Climate Change Assessment Report 4 (IPCC AR4) indicated that flood is likely to be a major cause of regional concern under warmer climates (Solomon et al. 2007). Floods accounted for 40% of the total number of natural disasters that occurred between 1985 and 2009, and they resulted in massive destruction in terms of economic loss and persons affected (Ferreira et al. 2011) (see Table 1.1). During the last decade of the twentieth century, floods killed 100,000 people and affected 1.4 billion people (Jonkman 2005). On average, floods affected 99 million people every year from 2000 to 2008 (Johnson 2010). Floods are currently one of the greatest threats to social security and sustainable development, and it is estimated that floods affect around 20–300 million people every year (Hirabayashi and Kanae 2009).

A.M. Dewan, *Floods in a Megacity: Geospatial Techniques in Assessing Hazards, Risk and Vulnerability*, Springer Geography, DOI 10.1007/978-94-007-5875-9_1,

Table 1.1 Number of reported disasters and humans affected, 1985–2009

Disaster type	Number of events (%)	People killed (%)	People affected (m) (%)
Floods	2,893 (40)	175,453 (13)	2,677 (53)
Storms	2,251 (31)	414,425 (31)	722 (14)
Extreme temperature	339 (5)	101,638 (8)	92 (2)
Earthquakes	656 (9)	601,032 (45)	136 (3)
Droughts	352 (5)	7,512 (1)	1,425 (28)
Other	829 (11)	47,825 (4)	16 (0)

Source: Ferreira et al. (2011)

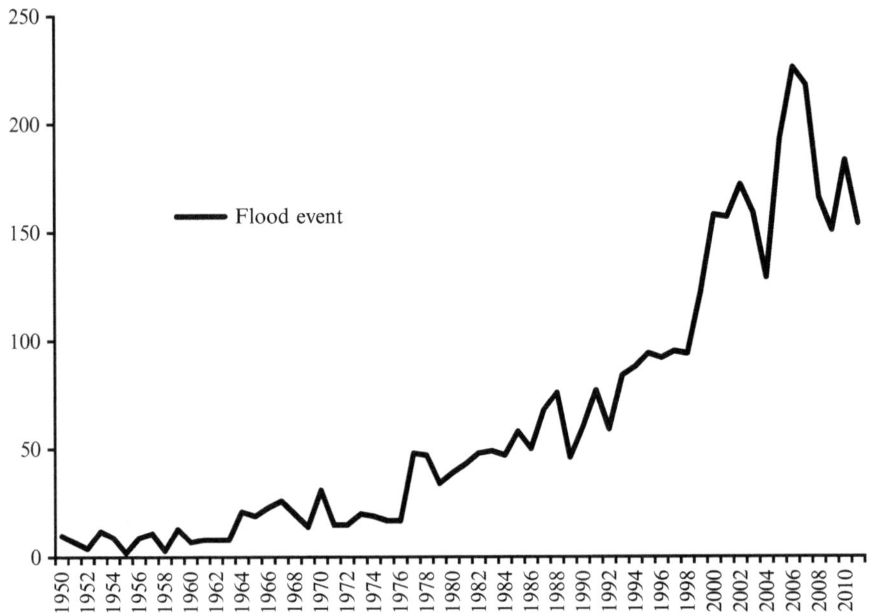

Fig. 1.1 Number of reported flood events between 1950 and 2011 (Source: EM-DAT/CRED, v. 12.07)

Figure 1.1 shows the trend of flood events around the world since 1950. Data from EM-DAT/CRED (v. 12.07) showed that 3,954 flood events (out of 7,849 hydrometeorological events) occurred between 1950 and 2011, of which 52.2% occurred during 2000–2011. In addition, the data showed that only 2% occurred during 1950–1959, 3.9% during 1960–1969, 6.6% during 1970–1979, 13.2% during 1980–1989, and 21.9% during 1990–1999.

Although fatalities from floods have declined considerably around the world, economic losses have become more pronounced, causing enormous monetary losses

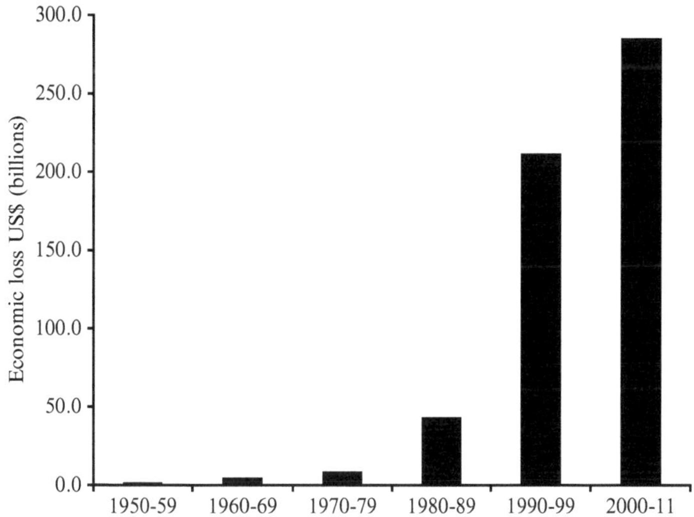

Fig. 1.2 Economic losses from floods, 1950–2011 (Source: EM-DAT/CRED, v. 12.07)

(see Fig. 1.2). Floods that occurred during 2000–2011 resulted in an estimated loss of more than US$285 billion. Likewise, the economic damage was more than $211 billion during 1990–1999. In contrast, flood-related losses were $1.8 billion during 1950–1959, $4.9 billion during 1960–1969, $8.8 billion during 1970–1979, and $43 billion during 1980–1989. However, the economic losses are disproportionate among continents. For example, Asia experienced the highest economic losses from 1950 to 2011, amounting to more than 60% of the global damage. This was followed by Europe (19%), the Americas (16.8%), Oceania (2.5%), and Africa (1.2%).

The most irreversible effect of flood is the loss of human lives, which is significantly higher in developing countries. Between 1950 and 2011, floods killed 2.3 million people and affected 3.5 billion people around the world, including more than two million during the 1950s and 68,361 during 2000–2011. Estimates show that over 95% deaths attributed to large floods occurred in developing countries, despite fewer large flood events (Ferreira et al. 2011). Of the deaths attributed to floods between 1950 and 2011, 96% occurred in Asia, 2.6% in the Americas, 0.9% in Africa, 0.4% in Europe, and 0.02% in Oceania (see Table 1.2). During 2000–2011, 68.8% of flood-related deaths occurred in Asia, followed by the Americas (15.7%).

Studies demonstrated that floods are expected to bring significant levels of misery in the coming years as a result of global climatic change (Whitfield 2012; Mirza 2011; Bouwer 2011; Pall et al. 2011; Hirabayashi et al. 2008; IPCC 2007a; van Aalast 2006; Kundzewich et al. 2010, 2005; Milly et al. 2002; Wetherald and Manabe 2002). A recent modeling study on climate change and flood probabilities suggested that up to 20% of the world's population is at risk of increased flooding due to climate warming (Kleinen and Petschel-Held 2007). This number is expected to increase with a further rise in global temperatures (Hirabayashi and Kanae 2009).

Table 1.2 Deaths from floods, by continent, 1950–2011

Continent	Persons killed	Percent (%)
Asia	2,268,968	96.13
Europe	7,846	0.33
Americas	61,857	2.62
Africa	21,134	0.90
Oceania	463	0.02
Total	**2,360,268**	**100.00**

Source: EM-DAT/CRED, v. 12.07

However, a general consensus is that poor countries will be disproportionately affected by water-related disasters that are driven by climate change because of their high rates of population and poverty and their poor adaptive capacity (Tol 2008; Adger 2006; Senga 2004).

South Asia, one of the most impoverished regions in the world, is at a high risk of flooding for many reasons (Kale 2003, 2012; Mirza 2011; Varis et al. 2011; Osti et al. 2011; Kumar et al. 2010; Gupta and Chakrapani 2007; Chowdhury and Ward 2007; Ferdous and Hossain 2005; Ahmad and Ahmad 2003; Chowdhury 2003a; Mirza 2003; Mirza et al. 2003; Anon 1993). With South Asia's extreme population density and rampant poverty, the problem of flooding is likely to exacerbate with climate warming, as intense precipitation is projected to swell (Solomon et al. 2007; Cruz et al. 2007; Milly et al. 2002; Palmer and Rälsänen 2002). Overall, the number of flood events in the region is increasing, with an average occurrence of nine per year, but, the peak discharges of major rivers have not been changed noticeably (Mirza 2003). The spatial distribution of floods in the region during 1950–2011 shows that India has the highest incidence of floods, followed by Bangladesh (see Fig. 1.3). Between 1950 and 2011, 0.14 million deaths were attributed to floods, while 1.2 billion people were affected in South Asia (see Table 1.3). Losses from floods in the region totaled $65.3 billion between 1950 and 2011, of which India had the highest economic damage (54.9%), followed by Pakistan (22.9%) and Bangladesh (18.5%).

1.2 Floods in Bangladesh

It is worth investigating the salient features of floods and flood management, as this is one of the costliest natural hazards in Bangladesh, resulting in estimated losses of US$175 million per year (Mirza 2011). This situation can be compared with the *ratchet effect* of vulnerability (Chambers 1989 in Pelling 2003), where each successive event diminishes the capacity of a group or an individual to withstand or recover from a later event (Pelling 2003). As a result, every year, Bangladesh must use a significant amount of resources to recover from floods, which reins in its economic progress.

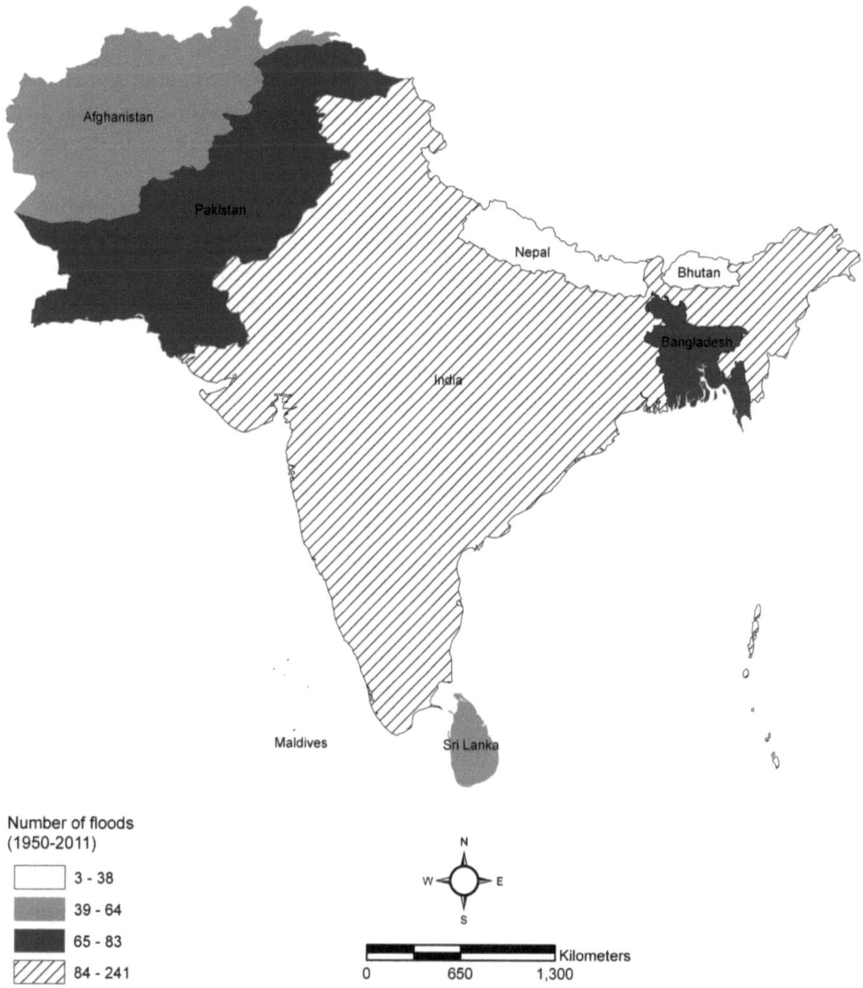

Fig. 1.3 Spatial distribution of flooding in South Asia, 1950–2011 (Source: EM-DAT, v. 12.07)

Table 1.3 Effects of floods on humans and the economy in South Asia, by country, 1950–2011

Country	Persons killed	Persons affected	Economic loss (US$'000)
Afghanistan	3,841	1,191,841	396,000
Bangladesh	52,102	312,232,241	12,038,400
Bhutan	222	1,600	0
India	60,672	811,179,289	35,802,188
Maldives	0	1,949	6,000
Nepal	6,183	3,581,347	1,037,242
Pakistan	15,375	68,653,534	14,968,178
Sri Lanka	1,162	12,636,752	979,364
Total	**139,557**	**1,209,478,553**	**65,227,372**

Source: EM-DAT/CRED, v. 12.07

Bangladesh is located in the downstream part of the Ganges–Brahmaputra–Meghna basin and is one of Asia's most at-risk countries of flooding (Osti et al. 2011; Islam 2006; Rahman et al. 2005; Jakobsen et al. 2005; Hossain 2004; Siddiqui and Hossain 2004; Mirza 2002; WARPO 2000; DMB 1998; Kibria 1970). Although fewer lives have been lost in the past few events, flood remains an inordinate threat to the people of Bangladesh.

Floods can be categorized into a number of types in Bangladesh depending on their nature, physical damage, recurrence interval, economic severity, and the extent of inundation. Based on the nature and types of floods, there are four categories: riverine, coastal, rain-induced, and flash flood (WARPO 2000). Riverine flood is the most common type and mainly originates from the spilling of waters from rivers that crisscross the country. Coastal flooding is accompanied by tropical cyclones and storm surges and is typically confined to the coastal belt. Rain-induced flooding results from intense precipitation and is widespread across the country. Flash flooding is characterized by an instantaneous rise in the water level and velocity (Chowdhury and Salehin 1987), and it is common in mountainous border districts.

Based on the extent of inundation and the degree of loss, floods can be divided into *normal* and *abnormal* types (Khalequzzaman 1994; Paul 1984). A normal flood, which inundates 20–25% of the land every year, is seen as highly beneficial to people and ecosystems (Islam 2001; Paul 1984, 1997; Haque and Zaman 1993; Brammer 1990a, b). In contrast, abnormal floods occur roughly once every decade (Dasgupta 2007; Mirza 2002) and are liable to submerge more than 50% of the total area (Mirza 1997).

Depending on timing, duration, and magnitude, abnormal floods are subdivided into four classes: early, late, prolonged, and high magnitude (Rasid and Mallik 1995). Floods that affect more than 10 million people are termed normal, while floods that inflict more than 30 million people are referred to as abnormal (Dasgupta 2007). Abnormal floods cause colossal damage to property and considerable suffering (FAO 2004; Islam and Sado 2000a, b; Khalequzzaman 1994). In addition to these, urban flooding has recently emerged as a serious threat to people and property in the major cities of Bangladesh (Islam 2006).

There have been many abnormal floods in Bangladesh between 1950 and 2010 (Hoque et al. 2011; Haque 2008; Rasheed 2008; Hofer and Messerli 2006; Rahman et al. 2005; Mirza 2003; Paul 1997; Karim 1995), and the floods of 1988, 1998, 2004, and 2007 were truly catastrophic (Irfanullah et al. 2011; Islam et al. 2010; Rahman et al. 2005; Mirza et al. 2003; DMB 1998; Ahmed 1989). A comparative analysis of the effects of these floods is presented in Table 1.4. Of these four events, the 1998 flood was unprecedented in terms of scale and intensity (del Ninno et al. 2002; Khan 1999); it affected 63 million people and claimed 1,050 lives (Dewan et al. 2003; Ali et al. 2002; Choudury 1998; DMB 1998), and the estimated economic loss was US$2 billion (Mirza et al. 2003; DMB 1998). The country experienced two major floods in 2004 and 2007, and the 2004 flood surpassed the estimated losses of the 1998 flood (see Table 1.4). Although the 2004 and 2007 floods inundated a relatively smaller portion of the country compared to the floods of 1988 and 1998, there was less time between the occurrences of the floods. Despite uncertainty in flood-related

Table 1.4 Effects of floods on Bangladesh during *abnormal* events[a]

Description	Flood year 1988	1998	2004	2007
Total inundated area (%)	61	68	38	42
People affected (million)	45	31	36	14
Total deaths (people)	2300	1100	750	1110
Livestock killed (number)	172,000	2,613,000	14,984	40,700
Roads damaged (km)	13,000	23,536	14,094	31,833
Houses damaged (fully/partially) (million)	7.2	0.98	4.00	1.1
Total loss US$ (billion)	1.4	2.0	2.3	1.4
Crop damaged (million metric tons)	2.12	2.8	1.3	2.1
Rice production loss (million tons)	1.65	2.06	1.00	1.2

[a]Compiled from various sources

projections, this reduced time can arguably be attributed to global warming (Kundzewich et al. 2010). However, research over the past two decades suggests that the risk of riverine floods could intensify in South Asia in response to a climatic shift, with major implications for future inundations in Bangladesh (Mirza 2011). A 23–29% increase in flooded areas is predicted for Bangladesh if global temperatures rise by 2°C (Mirza et al. 2003).

A mixture of physical and anthropogenic factors is accountable for exacerbating flooding in Bangladesh (Haque 2008; Rasheed 2008; Hofer and Messerli 2006; Mirza 2002; Rana and Lambert 2000; Karim 1995; Khalequzzaman 1994; Khalil et al. 1995; Reavill and Rahman 1995; Kubo 1993; Khalil 1990; Rob 1990; Oya 1990; Ali et al. 1989; Shailo 1988). For example, due to its geographic location, Bangladesh must drain significant amounts of transboundary runoff together with its own runoff generated from monsoonal precipitation during July–September (Mirza 2003). Increasing deforestation in the upstream region is also blamed for escalating flood damage (Ali 2007; Haque and Zaman 1993; BWDB 1987), although this is contentious (Hofer and Messerli 2006; Ives 1991; Hamilton 1987). In addition, flat topography (CBJET 1991), synchronization of major rivers' peak discharge (Brammer 1990a; Islam 2006), riverbed aggradation (Alexander 1989), the influence of El Niño–Southern Oscillation (Chowdhury 2003a), excessive urbanization and agricultural intensification (Khalequzzaman 1994), and pervasive poverty (Shimi et al. 2010) are other factors contributing to flood severity. Although recent floods are not physically different from past floods (Mirza 2011; Rogers et al. 1989), many studies demonstrated that increasing human activities and the gradual encroachment of floodplains are intensifying flood hazards from the national to the local level (Dewan and Yamaguchi 2008; Islam 2006; Rahman et al. 2005; Islam 2001; Haque and Zaman 1993). The data in Table 1.4 suggest that flood damage associated with moderate flooding has increased considerably. This can largely be attributed to environmental modifications driven by socioeconomic development. For example, 38%

areas were inundated during the 2004 flood; however, the economic losses were estimated to be around US$2.3 billion, which was higher than the worst flood on record in 1998.

To save lives and property from recurrent floods, there has been a steady growth of flood-control and drainage projects in Bangladesh since the 1960s (Cook 2010; Hossain 2004; Islam 2001; Alexander et al. 1998; Sultana and Thompson 1997; Mirza and Ericksen 1996; Haque and Zaman 1994, 1993). To date, 12,850 km of embankments (including 7,500 km of embankments along the major rivers), 25,580 km of drainage channels, and 4,190 sluices/regulators have been constructed across the country (Khan 2008; WARPO 2000), constituting around 30% of the country's total land area (Halls et al. 2008). Every year, Bangladesh spends about 20% of its national budget on water development projects (Islam 2001), which signifies that flood management is one of the country's top priorities. Based on the intensive appraisal of government and academic literature on flood management, Cook (2010) reported that academic literature is primarily *victim-centered*, while government studies are inclined to use engineering tools as a vehicle for flood management and do not focus on reducing damage by paying more attention to living with nature (White 1974; Myers and White 1993). As a result, controlling nature with structural measures has played a central role since the early time of flood management in Bangladesh.

In the context of a riverine country such as Bangladesh, where people have developed strategies over the millennia to cope with floods, structural solutions have been criticized for many reasons (Hutton and Haque 2004; Huq 1999; Paul 1997; Haque and Zaman 1994; Sklar 1993; Adnan 1991; Mirza 1991; Rashid 1991). Criticism related to engineering solutions became more pronounced after the adoption of the Flood Action Plan (FAP) (Jacobs and Wescoat 1994; Haque and Zaman 1994; 1993; Custers 1992; Alam 1991; Bingham 1989). It is worth noting that the FAP comprises 26 original components and a few supplementary activities, including both structural and nonstructural measures (Brammer 2004; Chowdhury 2000). Examining the perspective of flood management, solutions have chiefly been governed by engineering for many reasons, including technology and fund constraints. Owing to such limitations, Bangladesh had little choice but to comply with the suggestions made by collaborators, development partners, and experts in the country. As Dodson (1996, p. 184) argues, '*today, much of the funding and planning of flood control in Bangladesh is in the hands of foreign agencies*'. For instance, the multibillion dollar FAP project was not ratified by a legitimate parliament (Hughes et al. 1994). As a result, the country shows a disposition to control nature rather than work with it. Another factor that might influence the preference for a technical fix (engineering) was argued by Barua and van Ast (2011, p. 710):

> *'In general, physical flood control measures are emphasized more than non-structural measures. This has much to do with the financial implications: construction projects create huge streams of money that can be divided between the many actors involved'.*

Since flood-control works require large operating costs, Bangladesh can hardly bear the expense associated with regular maintenance of physical structures (Brammer 2010);

thus, weak performances of these structures during abnormal floods are frequently reported (Saleh and Mondal 1999; Thompson 1996; Hoque and Siddique 1995; Islam 1991; Zahurul 1991). A long-term study on the performance of flood-control works revealed that internal flooding became severe in protected areas due to hydraulic leakage (Bala et al. 2009; Faisal et al. 2003; Chowdhury et al. 1998). In addition, a significant increase in flood-hazard areas linked with embankments was also reported (WARPO 2000). Surprisingly, the duration of floods in many locations has also increased markedly in relation to flood-control works, which again warn of the efficacy of these structures (Rahman et al. 2005). In line with earlier studies (Paul 1984; Haque and Zaman 1993; Rasid and Paul 1987), Dasgupta (2007) reported that normal floods are indispensable and beneficial for Bangladesh. The study further corroborated that the embankments are unable to secure people and property during catastrophic floods. Additionally, breaching and/or overtopping of levees during floods is common, which places lives and property at an increased risk of flooding. Further, these flood-control works are based on a stationary climate; as stationarity is already dead, the existing structural measures are no longer valid (Milly et al. 2008). Consequently, the traditional concept of return periods is dead, and new research is needed in order to include uncertain information in design codes (Kundzewich et al. 2010).

Despite massive expenditure on flood defense, flood losses continue to rise, as evidenced by many recent studies (Mirza 1984, 2003, 2011; Hoque et al. 2011; Dewan et al. 2006; Rahman et al. 2005; Mahmud 2004; Haque and Zaman 1994). While *technocratic* solutions have clearly contributed to flood mitigation (Khan 2008), fish quantities, one of the vital sources of protein for Bangladeshi people, have decreased considerably (Halls et al. 2008; Alexander et al. 1998; Mirza and Ericksen 1996). The 217-km Brahmaputra Right Bank Protection Embankment is blamed for the westward shifting of the Brahmaputra–Jamuna channel (Haque and Zaman 1993, 1994). Damage associated with floods in protected areas is worse than in unprotected areas (Chowdhury 2003b; Chowdhury and Sato 2000; Stewart 1988). In addition, the deterioration of surface-water quality (Alam et al. 1996; Rasid and Mallik 1996), the degradation of soil quality (Alam and Samsuddin 1988), the increase of water- and vector-borne diseases (Emch 2000; Minkin et al. 1996), and the substantial reduction of wetlands (Nakashima and Khan 1994) were found to be the outcomes of massive structural schemes. The changes in river channels due to river engineering often lead to a considerable increase in flood risk (Arnaud-Fassetta 2003), which is already visible in Bangladesh.

As noted by Gillis (1993), continual attempts to manipulate nature could result in serious deterioration of the environment, which may be irreversible. It is now accepted that protective measures engender a false sense of security, which is often counterproductive (Gao et al. 2007; Few 2003; Etkin 1999; Green et al. 1991). For example, the construction of embankments encourages formal and informal developments of floodplains, which allow people to settle in hazardous locations (Pelling 2003; Haque 1994), thus giving rise to long-term vulnerability to natural hazards (Etkin 1999). Damage could occur in the event of flood overtopping or levee failure that is more severe than if the levee had never been built (Larson 1994 in Birkland et al. 2003).

1.3 Vulnerability of Megacities to Natural Hazards

Urbanization, fuelled by population growth and economic development, has two opposing facets. While megacities act as engines of economic and social improvement (Johnson 2010; Girard et al. 2003), they have become the most at-risk areas for environmental hazards, particularly in developing countries (International Federation of Red Cross and Red Crescent Societies [IFRCRCS] 2010; Kraas 2008; Munich Re 2004; Benson and Clay 2003; Mitchell 1999a; Anderson 1992; Davis 1987; Havlick 1986). For instance, the average number of victims of natural hazards is 150 times greater in developing countries, and economic losses are about 20 times higher (Wenzel et al. 2007). This clearly indicates the vulnerability of cities in developing countries to natural hazards.

A megacity is generally defined as an urban agglomeration that houses ten million or more inhabitants (United Nations [UN] 2003), although it has been defined differently (Kraas 2007). Numerous studies have documented a range of factors that influence the vulnerability of cities and megacities to natural hazards (Balica et al. 2012; Jha et al. 2012; Lall and Deichmann 2012; Hochrainer and Mechler 2011; World Bank 2011a; Lankao and Qin 2011; Finch et al. 2010; Bhattarai and Conway 2010; Showalter and Lu 2010; Adikari et al. 2010; Taubenböck et al. 2008, 2009; Tran et al. 2009a; Kubal et al. 2009; Meyer et al. 2009; Wenzel et al. 2007; Rashed et al. 2007; Sherbinin et al. 2007; Aragon-Durand 2007; Azar and Rain 2007; Cutter et al. 2003; Rashed and Weeks 2003; Montz and Tobin 2003; Quarantelli 2003; Pelling 2002, 2003; Wisner 2003; Godschalk 2003; Cross 2001; Moore 2001; Sanderson 2000; Mitchell 1993, 1999a, b; Uitto 1998; Hamza and Zetter 1998; Hewitt 1997; Steedman 1995; Horlick-Jones 1995; Kelly 1995; Solway 1994, 1999; Anderson 1992; Jones and Kandel 1992; Liverman 1986; Davis 1987; Havlick 1986). These studies deduce urban vulnerability to environmental hazards from diverse perspectives such as human–environment interactions, political economy or political ecology, as well as the increasing vulnerability of megacities to natural hazards resulting from transformations in population size, development patterns, and socioeconomic characteristics. As the socioeconomic systems of megacities—particularly in developing countries—are extremely dynamic (Kraas 2007), the interaction of global and local environmental processes exacerbates the risk of natural hazards (IPCC 2007b). Therefore, the effects of global climatic change driven by human activities are increasingly being investigated, indicating that multilevel drivers are accountable for the enhancement of urban vulnerability to natural hazards, predominantly in low- and middle-income countries (Hunt and Watkiss 2011; Parnell et al. 2007). To reduce the vulnerability of megacities to environmental hazards such as floods, it is imperative to recognize the nature and linkage of the key drivers involved.

The geography of many megacities—particularly their physical locations—makes them vulnerable to natural hazards. Historically, many cities have been built in the proximity of rivers, coasts, or fertile volcanic soils to take advantage of accessibility and benefits from natural endowments (Lall and Deichmann 2012; Godschalk 2003). In the

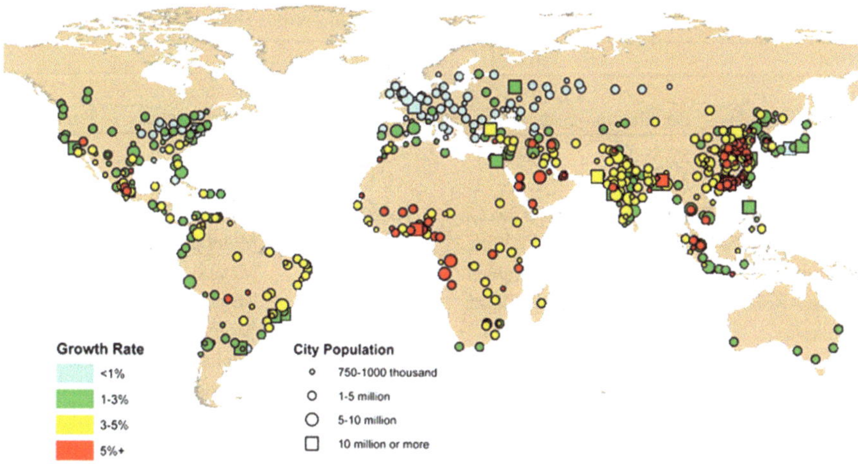

Fig. 1.4 Growth rates of urban agglomerations, 1970–2011 (Adapted from UN 2012)

early stages of development, these cities faced a relatively low risk of environmental hazards due to smaller population sizes and a lower concentration of infrastructures. However, with the proliferation of economies, larger populations, and a denser concentration of infrastructures, these places have become more hazardous as changes in social, economic, and the built environment profoundly alter landscape patterns (Cutter and Finch 2008), particularly in the cities of the global south (Kraas 2008). These changes give considerable rise to the exposure of natural hazards.

One of the greatest challenges that the world currently faces is the rapid growth of population in urban areas, particularly in developing countries. Estimates show that the world's urban population is expected to increase from 3.6 billion in 2011 to 6.3 billion in 2050. The largest increase will be in the cities of developing countries, where the population is projected to rise from 2.7 billion in 2011 to 5.1 billion in 2050 (UN 2012). In contrast, the population in rural areas is predicted to rise to 3.4 billion in 2021 and then decline gradually to 3.05 billion in 2050. Figure 1.4 shows the spatial distribution of the growth rates of urban agglomerations and city populations greater than 750,000 between 1970 and 2011.

While there were two megacities in the world in 1970, the number has since increased to 23 and is expected to increase to 37 in 2025, and many will be located in developing countries (Seto et al. 2010). The existing 23 megacities accommodate 9.9% of the world's population (UN 2012). Table 1.5 illustrates the temporal distribution of the 10 largest megacities since 1990 and shows that population growth in megacities in developed countries is nearly steady compared to developing countries. For instance, Dhaka was not in the list of top 10 megacities in 1990, whereas it currently ranks ninth and is likely to replace Beijing by 2025. This explosive increase of megacities is governed by many factors, including their greater contribution to the world's gross domestic product (GDP). It is estimated that around 97%

Table 1.5 Temporal distribution of population in top 10 megacities since 1990 (millions)

1990		2011		2025	
Urban agglomeration	Population	Urban agglomeration	Population	Urban agglomeration	Population
Tokyo	32.5	Tokyo	37.2	Tokyo	38.7
New York	16.1	Delhi	22.7	Delhi	32.9
Mexico City	15.3	Mexico City	20.4	Shanghai	28.4
Sao Paulo	14.8	New York	20.4	Mumbai	26.6
Mumbai	12.4	Shanghai	20.2	Mexico City	24.6
Osaka–Kobe	11.0	Sao Paulo	19.9	New York	23.6
Kolkata	10.9	Mumbai	19.7	Sao Paulo	23.2
Los Angeles	10.9	Beijing	15.6	Dhaka	22.9
Seoul	10.5	Dhaka	15.4	Beijing	22.6
Buenos Aires	10.5	Kolkata	14.4	Karachi	20.2

Source: UN (2012)

of the world's GDP comes from cities, which implies that the largest cities are also the largest economies (Johnson 2010). As urban areas provide more economic opportunities and other benefits, the low probability of extreme events occurring does not deter people from settling in risky areas. This is further corroborated by the fact that populations have not declined in well-known earthquake hot spots such as Istanbul, Tehran, and San Francisco (Lall and Deichmann 2012). Many people contended that disasters in megacities could result in substantial economic losses, which have profound effects on both local and global economies (Mitchell 1999b; Uitto 1998; Nicholls 1995).

While urban areas are not disaster-prone, per se (Srinivas et al. 2009), many argue that urban sprawl accompanied by rapid population growth is one of the main causes of the increasing vulnerability of megacities to environmental risks (Dewan et al. 2012; Tran et al. 2009a; Rabbani 2009; Szlafsztein and Sterr 2007; Pelling 2003; Mitchell 1999b; Steedman 1995; Hewitt 1997; Havlick 1986). For instance, piecemeal approaches to urban developments originating from inappropriate planning and legislation can deepen vulnerability (Pelling 2003), although others have differing opinion on this issue (Taubenböck et al. 2009; Coleman and Schofield 1986). However, structural processes that accelerate urbanization and population concentrations can significantly increase disaster vulnerability of megacities, predominantly in poor countries (Parnell et al. 2007; Hamza and Zetter 1998).

Indeed, the vulnerability of urban areas increases as cities continue to expand into areas such as wetlands, swamps, and floodplains. For example, the continual encroachment of floodplains can lead to appalling conditions (e.g., reduction in flood conveyance routes), and in the event of a disaster, the effects could be far greater than for previous disasters (Quarantelli 2003; White et al. 2001; Haque 1994). For example, artificial drainage systems may be an alternative to natural water bodies and swamps if adequate drainage facilities are provided to ensure the conveyance of massive volumes of floodwater. However, this is evidently absent in many fast-growing cities (Barua and van Ast 2011; Stalenberg and Vrijling 2009;

Sherbinin et al. 2007; Gupta 2007; Alam and Rabbani 2007), which amplifies flood risk. In contrast, rapid urbanization with few provisions of safe sanitation and drinking water could result in the spread of water-borne diseases during and after the event, which is already apparent in many cities (Ahern et al. 2005; Bull-Kamanga et al. 2003; Kunii et al. 2002) and poses serious health risks to urban dwellers (Few 2003; Tapsell et al. 2002).

Due to the growing demand for land for housing and industry (Lall and Deichmann 2012; Gupta 1994), structures built on floodplains are highly susceptible to environmental risks, and in the event of a disaster, they are the first to be affected (Quarantelli 2003). For instance, structures built on alluvial plains evidently suffered incredible losses during the Adana–Ceyhan earthquake in Turkey in 1998 (Ulusay et al. 2002). In addition, poor building materials and flaws in the construction process could amplify the vulnerability of urban settlements (Ansary 2003; Solway 1994), which may result in the loss of lives and property. Therefore, the quality of housing is an important parameter that significantly influences urban vulnerability to disasters caused by natural hazards (Tipple 2006). For instance, earthquakes of nearly equal magnitudes that occurred in Haiti and Chile in 2010 resulted in 200,000 deaths and more than one million homeless in Haiti, whereas a slightly higher magnitude of earthquake in Chile killed 100 people. Aside from the differences in disaster preparedness, the quality of housing was the most dominant factor that contributed to the higher death toll and severe damage in Haiti (Satterthwaite 2010).

The vulnerability of megacities is further exacerbated by the rise of rampant poverty; the link between poverty and disaster risk is clearly visible (Sanderson 2012, 2000; Satterthwaite 2010; Rayhan 2010; Few 2003; Gupta 1994; Burton et al. 1993). A recent estimate shows that the number of slum dwellers in the developing world has increased from 776.7 million in 2000 to about 827.6 million in 2010 (UN–HABITAT 2010). Due to a lack of access to resources, slum dwellers often make their shelters in hazardous areas and environmentally degraded places of the city with little access to quality services such as housing, potable water, and safe sanitation (Gilbert 2005). These deplorable conditions make them more exposed to the risk of natural hazards. Moreover, because of their *illegal* status, they are often excluded from the planning process (Quarantelli 2003), which makes these socially disadvantaged people most vulnerable to natural hazards (Braun and Aßheuer 2011; Jabeen et al. 2010; Chatterjee 2010; Nchito 2007; Cutter et al. 2003; Thomas and Mitchell 2001; Rashid 2000; Chan and Parker 1996). Taking examples from many megacities around the world, Chatterjee (2010) states that the marginal sections of urban society are not included in formal hazard mitigation preparation, which is essential to achieve successful hazard mitigation strategies (Quarantelli 2003; Davis 1987).

Globalization that extends local and global economies reportedly affects the incidence of disasters in urban areas, particularly in large cities (Benson and Clay 2003; Albala-Bertrand 2003).

Disaster awareness and the differences in people's risk perception may have a significant influence on the vulnerability of urban communities. The development of hazard-prone areas or building protection measures by structural fixes could change the risk behavior of people, which could have tremendous effects on long-term vulnerability (Etkin 1999). In doing so, it could substantially enhance the risk of natural hazards of a society (Gao et al. 2007). People's disbelief of the early warning system may also result in greater vulnerability to natural hazards if the communication of risks is not properly conversed (Haque and Blair 1992).

In addition, complex and aging infrastructures, lack of critical facilities, the power structure of the city, and weaknesses in response and relief capabilities escalate the vulnerability of megacities to natural hazards (Ranger et al. 2011; Taubenböck et al. 2009; Kraas 2008; Wenzel et al. 2007; Quarantelli 2003; Pelling 2003). Many megacities in developing countries seriously lack accurate information regarding past disasters (Johnson 2010); thus, it is difficult to develop suitable countermeasures. For example, the megacity of Dhaka experienced four large floods in the recent past; however, it does not have precise information regarding the extent of the floods or the damage and fatalities, which is a significant hurdle to predicting the likely pattern of risk from natural hazards (Huq 1999). Consequently, the existing disaster management systems of emerging cities is increasingly being tested, and the governments of these countries consistently struggle to allocate adequate resources to reduce their vulnerability to natural hazards (Mitchell 1999a). Due to a number of complexities linked with megacities, it is a challenging job to ameliorate the risk of natural hazards (Wisner 2003). However, opportunities exist to develop improved response systems and resilience if the socioeconomic components of these cities are incorporated into the vulnerability and risk assessments (Kraas 2007).

Historically, disaster research was mainly confined to rural areas; however, due to the increasing concentration of populations and assets (physical, social, and human) and the growing economic damage from natural hazards in urban areas, research on urban vulnerability has recently received ample interest worldwide (Mitchell 1993). Urban disaster research predominantly gained momentum and renewed interest when the UN-sponsored International Decade for Natural Disaster Reduction (1990–2000) gave high priority to the issue of megacities and disasters (Mitchell 1999a; Solway 1994). Afterward, two academic journals (*Applied Geography* and *GeoJournal*) published theme editions on urban disasters, but they were mostly dominated by examples from western cities. At the same time, *Built Environment*, another leading journal, published a theme issue on *Hazards in the Built Environment*.

The most recent international effort is the Hyogo Framework for Action 2005–2015, where each signatory state agreed to (1) integrate disaster prevention, mitigation, preparedness, and vulnerability measures into development practices for achieving sustainability; (2) enhance local capacity by using knowledge, education, and innovation for creating hazard resilience society; and (3) identify underlying risk factors and incorporate them into the design and implementation of disaster preparedness for effective response at all levels (United Nations International Strategy for Disaster Reduction [UNISDR] 2005). Currently, there is an increasing focus on the effects of climate change policy on the role of cities, as well as mitigation and adaptation to natural hazards (World Bank and UN-ISDR 2008).

As a result of these efforts, several urban vulnerability models have been proposed (Müller 2012; Taubenböck et al. 2008; Rashed and Weeks 2003; Cutter 1996;

Menoni and Pergalani 1996; Burton et al. 1993; Mitchell et al. 1989) and applied to few large cities such as Istanbul (Taubenböck et al. 2008), California (Rashed and Weeks 2003), South Carolina (Cutter et al. 2000), and Santiago (Müller 2012); however, most of them are based on developed countries, with the exception of Müller (2012). These models are primarily rooted in the concept of human–nature interaction, in which natural hazards are viewed as dynamic phenomena and humans are the agents of environmental modifications as well as the victims of hazards (Kates 1996). Nevertheless, urban disaster research is significantly lacking in developing countries, particularly in megacities, where most of the effects of natural hazards are strongly felt. Although an urban area is exposed to a variety of natural, technological, and man-made hazards, for simplicity, the following sections will largely concentrate on flood—particularly urban flooding—as dealing with other hazards is beyond the scope of this book.

1.3.1 Urban Flooding: A Global Phenomenon

Flooding used to be limited to rural areas, but it has now become pervasive in urban areas as well (Green 1994 in Pelling 2003). With their high population densities and concentration of diverse economic activities, cities around the world—particularly in developing countries—face tremendous challenges in dealing with flood-related problems (Montoya 2003). Table 1.6 presents the ten largest cities and their risk deciles by type of natural hazard. It reveals that many of the largest cities are exposed to more than one type of hazard. For example, the largest megacity in the world, Tokyo, is susceptible to cyclones and floods. Of the natural hazards analyzed, estimates show that at least 233 cities across the globe are at a high risk of flooding, potentially affecting 663 million people (UN 2012) (see Fig. 1.5). This report further demonstrates that flood is the most frequent natural disaster and continues to be a major threat for many cities around the world.

Floods in urban areas intensify with the increase of impervious surfaces such as roofs, roads, parking lots, and pavements (Kundzewich et al. 2010), which cause changes in runoff conveyance networks. Over the past few years, many cities around the world have experienced disastrous floods; notably, a flood in Bangkok, Thailand, lasted for 175 days, took 815 lives, and caused about US$45.7 billion in damage (World Bank 2011b). At least five major floods ravaged Queensland and Victoria in Australia during 2010–2011 (DFO 2012; Wilby and Keenan 2012). The Indian commercial capital, Mumbai, experienced an unprecedented flood in 2005 that resulted in estimated damages of around US$2 billion and 500 fatalities (Ranger et al. 2011; Gupta 2007). Floods in New Orleans, United States, in 2005 engulfed 80% of the city and resulted in enormous damage (Bizimana and Schilling 2010), which suggests that cities in the developed world are also not immune to floods.

Since the level of urbanization is expected to increase considerably—particularly in Asian megacities (UN 2012)—the degree of vulnerability to natural hazards is likely to increase exponentially (Sharma et al. 2011). A recent study indicated that,

Table 1.6 Top 10 largest city populations in 2011, at 8th–10th risk deciles by hazard type

	Cyclone	Drought	Earthquake	Flood	Landslide	Volcano
1	Tokyo, Japan	Kolkata, India	Los Angeles, USA	Tokyo, Japan	Taipei, China	Napoli, Italy
2	Shanghai, China	Karachi, Pakistan	Manila, Philippines	Delhi, India	Bandung, Indonesia	Quito, Ecuador
3	Manila, Philippines	Los Angeles, USA	Istanbul, Turkey	Mexico City, Mexico	Quito, Ecuador	Bogor, Indonesia
4	Osaka–Kobe, Japan	Chennai, India	Lima, Peru	New York, USA	San Salvador, El Salvador	Malang, Indonesia
5	Guangzhou, China	Lahore, Pakistan	Tehran, Iran	Shanghai, China	Kaohsiung, China	
6	Shenzhen, China	Ahmadabad, India	Santiago, Chile	Sao Paulo, Brazil	San Jose, Costa Rica	
7	Seoul, Republic of Korea	Santiago, Chile	San Francisco, USA	Dhaka, Bangladesh		
8	Dongguan, China	Belo Horizonte, Brazil	Kunming, China	Kolkata, India		
9	Hong Kong, China	Luanda, Angola	Nagoya, Japan	Buenos Aires, Argentina		
10	Foshan, China	Yangon, Myanmar	Izmir, Turkey	Rio de Janeiro, Brazil		

Source: UN (2012)

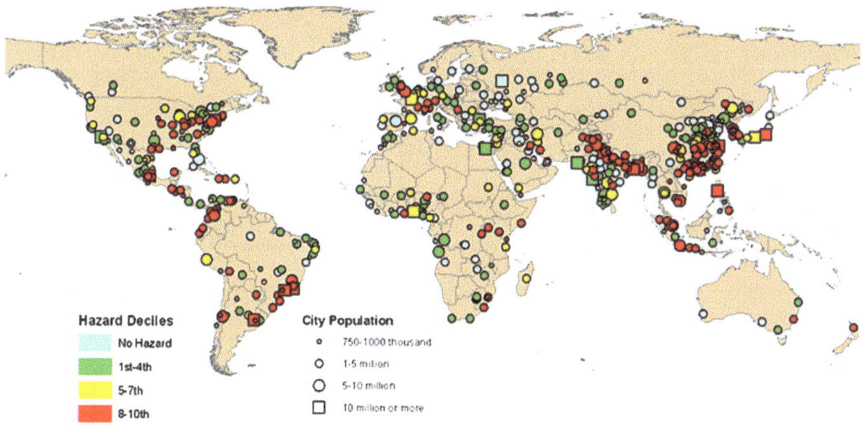

Fig. 1.5 Urban agglomerations by size class and potential risk of flooding, 2011 (Adapted from UN 2012)

due to weaknesses in the adopted disaster preparedness system, communities' capacity to cope with extreme events is being reduced in many Asian megacities (Osti et al. 2011), and it is expected to worsen. Further, the extreme dynamics of the socioeconomic system that result from global change, development practice, and political instability could lead to more frequent and disastrous floods if disasters risk management is not mainstreamed into the planning process of these Asian megacities (Adhikari et al. 2010).

1.3.2 Causes and Types of Urban Flooding

As noted earlier, urban flooding is a matter of apprehension for developed and developing countries because of its dramatic effects on people and infrastructures. Typically, urban flood results from a complex combination of different causes, including the spread of impermeable soils over rural lands. As the risk of urban flooding increases with the rapid conversion of floodplains to urban surfaces (Alkema 2003; Mitchell 1999a), it is important to briefly describe the phenomenon in order to understand the causes and types of urban flooding.

Urban flood mainly results from modifications of the hydrological cycle by land-use transformation. Land-use change accompanied by impervious surfaces can considerably affect flooding in urban areas, including timing, intensity, and the extent of inundation (Suriya and Mudgal 2012; Du et al. 2010; Dewan and Yamaguchi 2008; Nirupama and Simonovic 2007; Burns et al. 2005; Weng 2001; Paul and Meyer 2001; Leopold 1994; Hirsch et al. 1990). However, the hydrological responses vary according to different stages of urban growth (Kibler et al. 1981). For example, increasing sedimentation and decreasing evapotranspiration are common in the early stages of urban development, while the volume of runoff and flood damage potential is significantly

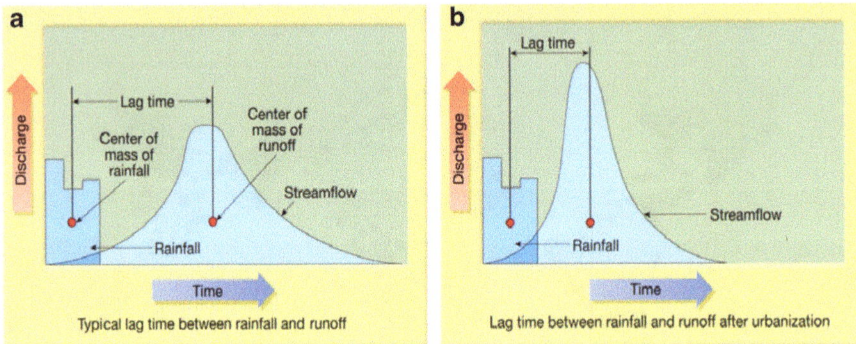

Fig. 1.6 Effects of urbanization on floods (Adapted from Tarbuck et al. 2005)

amplified with the spread of built-up areas (Weng 2001; Parker 1999). As vegetation is cleared and the imperviousness increases with hard materials and artificial surfaces, the amount of infiltration decreases. This affects the runoff and rainfall process in an urban watershed (Goudie 1990) and thus gives rise to flood frequency.

As seen in Fig. 1.6, in a natural setting, a significant time delay, called lag time, exists between the rainfall and the peak of the flood. However, with increased impermeable soils through urban development, the lag time dropped sharply, resulting in the rapid rise and fall of floods. This situation is further exacerbated by increased runoff, which quickly blocks storm drains and channels with sediment, waste, and debris—a situation analogous to a bathtub shower when the drain is partly blocked by soap (Keller and Blodgett 2008). Hence, the extent of flooding in an urban area depends on both the peak water level and the condition of the drainage system. Although channeling the streams is a popular means of ameliorating flood loss in urban areas, this in turn elevates the vulnerability of a community to floods by promoting urban and industrial growth within the floodplains (Zahran et al. 2008; Chang and Franczyk 2008). It also tempts people to settle in the floodplains, as land prices are usually lower and the local land-use control over floodplains is minimal. However, the effect of urbanization on floods is reduced with the increase of larger and more frequent floods (Hollis 1975).

Urban flooding can be categorized into a number of types. For example, Jha et al. (2012) categorized urban flood into five major types, while ActionAid (2006) classified four major types. However, both studies indicated that at least four types of floods are common in urban areas. They are:

(a) *Fluvial or river floods* occur when rivers flow over their banks. Due to the construction of flood-control works and floodplain development through land-use change, river waters can inundate urban areas when natural waterways lose their ability to drain massive amounts of floodwaters. A recent example is the flooding in New Orleans in 2005, which resulted from the breaching of 53 different flood levees (Bizimana and Schilling 2010). Asian cities, particularly those that are located in the proximity of rivers, are highly prone to fluvial flooding.

(b) *Pluvial or overland flood* is also known as localized flood and is mainly the outcome of increased runoff from intense rainfall that cannot be handled by the existing drainage systems. Currently, pluvial floods in many cities have become a common feature, especially in the areas where storm water management systems have not kept pace with urban development. Pluvial or overland floods can severely affect people by paralyzing the regular functioning of the urban system. For example, the 2007 floods in the Hull area in the UK were examples of pluvial floods that resulted from prolonged rainfall (Jha et al. 2012). Localized flooding is also very common in tropical cities, predominantly in the wet season.

(c) *Coastal floods* are common in cities that are located in the proximity of the coast. At least 85 cities in the world are at a high risk of coastal flooding (UN 2012). Typically, a cyclone accompanied by a storm surge can inundate urban areas with tragic consequences. However, tsunamis can also trigger coastal flooding in cities, ravaging large areas in a short time. The 2004 Indian Ocean tsunami, the 2011 Japanese tsunami, and the 1991 cyclone in Bangladesh are recent examples of coastal flooding that resulted in colossal damage and killed thousands of people. In addition, experts fear that coastal floods could be exacerbated due to rising sea levels.

(d) *Flash flood* is one of the most damaging phenomena for a megacity because of its sudden occurrence. Flash floods can originate in local convective storms and glacial lake outburst floods (GLOF). Due to the increase of impermeable surfaces, this flood rises very quickly and exceeds the capacity of storm water facilities to drain early, thus resulting in severe damage to urban infrastructures. The flash flood that occurred in Jeddah city, Saudi Arabia, in November 2009, killed 100 people and caused estimated losses of US$270 million (Jha et al. 2012).

It is now obvious that flooding is a major concern for urban sustainability in the world—primarily in cities that accommodate ten million or more inhabitants in developing nations. A comprehensive risk assessment for an urban area involving hazard and vulnerability analyses is therefore indispensable (World Bank 2011a) to identify existing shortcomings and distinguish the actions and resources that are required to lessen the probability of threatening events and their consequences (Tran et al. 2009b; United Nations Human Settlement Program [UNHSP] 2004).

Since people's ability to save lives and property from devastating floods is limited (Birkland et al. 2003), greater emphasis must be placed on prevention and mitigation practices (Kreimer and Munasinghe 1991). This can be accomplished with a number of tasks such as identifying actual hazards and risks areas (Jha et al. 2012; Meyer et al. 2009; Greiving et al. 2006; Islam and Sado 2000a, b, c; Paul 1997) and evaluating vulnerability (Kienberger 2012; Yoon 2012; Cutter et al. 2009; Barroca et al. 2006; Birkmann 2006; O'Brien et al. 2004; Cutter 1996). Other tasks include improving early warning systems (Rahman et al. 2005; Aziz et al. 2003; Paudyal 2002; Liong and Sivapragasam 2002; Liong et al. 2000), evacuation mapping and practice (Sanyal and Lu 2009; Dasgupta 2007; Chakraborty et al. 2005; Cova 1999), and promoting proper mitigation measures such as floodplain zoning (Birkland et al. 2003; Hossain 1998). Continued efforts should be made to work with nature; otherwise, structural measures could change individuals' or community responses to natural hazards (Birkland et al. 2003), which can ultimately increase the risk of disaster (Tobin and Montz 1997).

In recent years, a paradigm shift in flood policy has been evident across the world, that is, flood-risk management has become the focus rather than the traditional concept of flood protection (Büchele et al. 2006; Schanze 2006), which comprises flood-risk assessment (Meyer et al. 2009; Apel et al. 2009). The primary aim of risk assessment is to identify risk-related problems and select appropriate measures to reduce risks associated with hazards (Smith 2001). Hence, disaster risk reduction (DRM) has become an integral part of the entire risk management process and could assist in saving lives and property from catastrophic events (Bendimerad 2009). In developed countries, hazard estimation, mapping vulnerability, and risk communication are used to reduce risk and vulnerability in relation to natural hazards (Showalter and Lu 2010; Cutter et al. 2003; Odeh 2002; Morrow 1999; Mitchell 1989; Kasperson 1986); however, structural solutions are popular in many countries, particularly in Bangladesh.

Spatial models that allow the integration of diverse databases with advanced techniques (Taubenböck et al. 2008; Cutter 2003; Rashed and Weeks 2003; Weichselgartner 2001; Rosenfield 1994) provide ample opportunities to quantify and determine the vulnerability of megacities to natural hazards. Currently, detailed information on risks of, and vulnerability to, natural hazards is only available at the metropolitan level in developed countries (Johnson 2010), and most developing countries lack such information despite their increasing vulnerability to floods and other natural hazards.

Accurate information about past events and the spatial distribution of environmental risks is no less important in enabling the government and policy makers to develop appropriate countermeasures. Keeping this in mind, the major objective of this work is to develop spatial information on hazards, risks, and vulnerability relating to flooding for one of the fastest growing megacities in the world. Specifically, the basic premise of this work is to create a series of maps relating hazards, risks, and vulnerability, which can easily be interpreted to allow improved decision making for mitigating flood-related damage. Dhaka, which is a megacity in Bangladesh, is considered the focal point, as lack of information is exacerbating the city's vulnerability to natural hazards, particularly flood. Historical floods from 1988 to 2009 were considered in order to answer a number of research questions, including the extent of the floods in past events; the causes of vulnerability to floods; the scale of damage to human settlements such as housing, whether there is spatial variability in flood vulnerability; and the spatial distribution of flood risk. This information is expected to help ameliorate the increasing flood risk in Dhaka.

1.4 Outline of the Book

Chapter 2 presents a theoretical framework on hazards, risks, and vulnerability. Existing models on disaster vulnerability are discussed, as well as how geospatial techniques are being exploited to the mapping and quantifying of risk from floods.

On the basis of existing models and techniques within a geospatial framework, a conceptual model has been developed and is described in Chap. 2. Chapter 3 describes the environmental and social settings of the study area. Additionally, a historical account of previous floods, damage statistics, and the major causes of flood vulnerability is presented. Chapter 4 presents flood estimations and the accuracy of flood maps derived from satellite data for the historical floods of 1988–2009. In addition, synthetic depth-damage curves for different housing types and road networks are developed and discussed. The identification of hazardous flood areas is presented in Chap. 5, which considers two hydrologic parameters derived from multi-temporal spatial databases. Chapter 6 elucidates vulnerability and risk assessments using physical and social models. It also illustrates flood risk zoning at the community level in Dhaka. Finally, Chap. 7 presents the conclusions and contributions of this research to urban surface water flood hazard mitigation, as well as further research directions.

References

ActionAid (2006) Climate change, urban flooding and the rights of the urban poor in Africa. ActionAid International, Archway

Adger W (2006) Vulnerability. Glob Environ Chang 16(3):268–281

Adikari Y, Osti R, Noro T (2010) Flood-related disaster vulnerability: an impending crisis of mega-cities in Asia. J Flood Risk Manage 3(3):185–191

Adnan S (1991) Floods, people and environment: institutional aspects of flood protection programmes in Bangladesh. Research and Advisory Services, Dhaka

Ahern M, Kovats RS, Wilkinson P, Few R, Mattheis F (2005) Global health impacts of floods: epidemiological evidence. Epidemiol Rev 27(1):36–46

Ahmad QK, Ahmad AU (2003) Regional cooperation in flood management in the Ganges-Brahmaputra-Meghna region: Bangladesh perspective. Nat Hazard 28(1):191–198

Ahmed M (ed) (1989) Flood in Bangladesh. Community Development Library, Dhaka

Alam SMN (1991) Conquering nature: myth and the reality of flood control in Bangladesh. A paper presented at the UCLA international conference on the impact of natural disasters, Los Angeles, CA

Alam M, Rabbani MG (2007) Vulnerabilities and responses to climate change for Dhaka. Environ Urban 19(1):81–97

Alam MS, Samsuddin D (1988) Effect of flood control embankments on properties of soil. Jahangirnagar Rev (Part A) 12:29–34

Alam BN, Damole LN, Wickramanyake E (1996) Effects of flood mitigation measure: lessons from Dhaka flood protection project, Bangladesh. Asian Profile 24(6):511–524

Albala-Bertrand JM (2003) Urban disasters and globalization. In: Kreimer A, Arnold M, Carlin A (eds) Building safer cities: the futures of disaster risk. The World Bank, Washington, DC, pp 75–82

Alexander D (1989) Consequences of floods in developing countries: international perspectives for disaster management. In: Proceedings of the international seminar on Bangladesh floods: regional and global environmental perspectives, Dhaka, 4–6 March, p 11

Alexander MJ, Rashid MS, Shamsudding SD, Alam MS (1998) Flood control, drainage and irrigation projects in Bangladesh and their impacts on soils: an empirical study. Land Degrad Dev 9(3):233–246

Ali AMS (2007) September 2004 flood event in south western Bangladesh: a study of its nature, causes and human perception and adjustment to a new hazard. Nat Hazard 40(1):89–111

Ali A, Quadir DA, Huh OK (1989) Study of river flooding hydrology in Bangladesh with AVHRR data. Int J Remote Sens 10(12):1873–1891

Ali MA, Seraj SM, Ahmed S (eds) (2002) Engineering concerns of flood. Bangladesh University of Engineering and Technology, Dhaka

Alkema D (2003) Flood risk assessment for EIA: an example of a motorway near Trento, Italy. Acta Geologica 78:147–153

Anderson MB (1992) Metropolitan areas and disaster vulnerability: a consideration for developing countries. In: Kreimer A, Munasinghe M (eds) Environmental management and urban vulnerability. The World Bank, Washington, DC, pp 77–92

Anon (1993) Floods and drought could increase in South Asia. Bull Am Meteorol Soc 74(7):1400

Ansary MA (2003) Seismic risk in urban areas of Bangladesh. Ministry of Environment and Forest, Dhaka

Apel H, Aronica GT, Kreibich H, Theiken AH (2009) Flood risks analyses – how detailed do we need to be? Nat Hazard 49(1):79–98

Aragon-Durand F (2007) Urbanization and flood vulnerability in the peri-urban interface of Mexico City. Disasters 31(4):477–494

Arnaud-Fassetta G (2003) River channel changes in the Rhone Delta (France) since the end of the Little Ice Age: geomorphological adjustment to hydroclimatic change and natural resources management. Catena 51(2):141–172

Azar D, Rain D (2007) Identifying population vulnerable to hydrological hazards in San Juan, Puerto Rico. GeoJournal 69(1–2):23–43

Aziz F, Tripathi NK, Mark O, Kusanagi M (2003) Flood warning and evacuation system using MIKE-11 and GIS. Asian J Geoinf 3(3):31–39

Bala SK, Islam AKMS, Chowdhury JU, Rahman MR, Haque MA, Khan MSA, Salehin M (2009) Performance of flood control works around Dhaka city during major floods in Bangladesh. A paper presented in the 2nd international conference on water and flood management, Dhaka

Balica SF, Wright NG, van der Muelen F (2012) A flood vulnerability index for coastal cities and its use in assessing climate change impacts. Nat Hazard. doi:10.1007/s11069-012-0234-1

Bangladesh Water Development Board (BWDB) (1987) Flood in Bangladesh: investigation, review and recommendation for flood control. BWDB, Dhaka

Barroca B, Bernardara P, Mouchel JM, Hubert G (2006) Indicators for identification of urban flooding vulnerability. Nat Hazard Earth Syst Sci 6(4):553–561

Barua S, van Ast JA (2011) Towards interactive flood management in Dhaka, Bangladesh. Water Policy 13(5):693–716

Bendimerad F (2009) State-of-the-practice report on urban disaster risk management. Earthquakes and megacities initiative. Available at www.emi-megacities.org

Benson C, Clay EJ (2003) Disasters, vulnerability and the global economy. In: Kreimer A, Arnold M, Carlin A (eds) Building safer cities: the future of disaster risk. The World Bank, Washington, DC, pp 3–32

Bhattarai K, Conway D (2010) Urban vulnerabilities in the Kathmandu valley, Nepal: visualizations of human/hazard interactions. J Geogr Inf Syst 2(2):63–84

Bingham A (1989) Floods of aid for Bangladesh. New Sci 124(1693):42–46

Birkland TA, Burby RJ, Conrad D, Cortner H, Michener WK (2003) River ecology and flood hazard mitigation. Nat Hazard Rev 4(1):46–54

Birkmann J (ed) (2006) Measuring vulnerability to natural hazards: towards disaster resilient society. United Nations University Press, Tokyo

Bizimana JP, Schilling M (2010) Geo-information technology for infrastructural flood risk analysis in unplanned settlement: a case study of informal settlement flood risk in the Nyabugogo flood plain, Kigali city, Rwanda. In: Showalter PS, Lu Y (eds) Geospatial techniques in urban hazard and disaster analysis. Springer, Dordrecht, pp 99–124

Bouwer LM (2011) Have disaster losses increased due to anthropogenic climate change? Bull Am Meteorol Soc 92(1):39–46

Brammer H (1990a) Flood in Bangladesh: I geographical background to the 1987 and 1988 floods. Geogr J 156(1):12–22

Brammer H (1990b) Flood in Bangladesh: II flood mitigation and environmental aspects. Geogr J 156(2):158–165
Brammer H (2004) Can Bangladesh be protected from floods? The University Press Limited, Dhaka
Brammer H (2010) After the Bangladesh flood action plan: looking for the future. Environ Hazard 9(1):118–130
Braun B, Aßheuer T (2011) Floods in megacity environments: vulnerability and coping strategies of slum dwellers in Dhaka/Bangladesh. Nat Hazard 58(2):771–787
Büchele B, Kreibich H, Kron A, Thieken A, Ihringer J, Oberle P, Merz B, Nestmann F (2006) Flood-risk mapping: contribution towards an enhanced assessment of extreme events and associated risks. Nat Hazard Earth Syst Sci 6(4):485–503
Bull-Kamanga L, Diagne K, Lavell A, Leon E, Lerise F, MacGregor H, Maskrey A, Meshack M, Pelling M, Reid H, Satterthqaite D, Songsore J, Westgate K, Yitambe A (2003) From everyday hazards to disasters: the accumulation of risk in urban areas. Environ Urban 15(1):193–204
Burns D, Vitvar T, McDonnell J, Hassett J, Duncan J, Kendall C (2005) Effects of suburban development in runoff generation in the Croton river basin, New York, USA. J Hydrol 311(1–4):266–281
Burton I, Kates R, White GF (1993) The environment as hazard, 2nd edn. Guilford Press, New York
Chakraborty J, Montz BE, Tobin GA (2005) Population evacuation: assessing spatial vulnerability in geophysical risk and social vulnerability to natural hazards. Nat Hazard Rev 6(1):23–33
Chan NW, Parker DJ (1996) Response to dynamic flood hazard factors in peninsular Malaysia. Geogr J 162(3):313–325
Chang H, Franczyk J (2008) Climate change, land use change and floods: towards an integrated assessment. Geogr Compass 2(5):1549–1579
Chatterjee M (2010) Slum dwellers response to flooding events in the megacities of India. Mitig Adapt Strateg Global Chang 15(4):337–353
China-Bangladesh Joint Expert Team (CBJET) (1991) Study report on flood control and river training project on the Brahmaputra river in Bangladesh, vols 1 and 2, Dhaka
Choudury M (1998) Report on the Bangladesh flood 1998. Disasters Management Bureau, Dhaka
Chowdhury MR (2000) As assessment of flood forecasting in Bangladesh: the experience of the 1998 flood. Nat Hazard 22(2):139–163
Chowdhury MR (2003a) The El Nino-Southern Oscillation (ENSO) and seasonal flooding – Bangladesh. Theor Appl Climatol 76(1–2):105–124
Chowdhury MR (2003b) The impact of 'Greater Dhaka Flood Protection Project' (GDFPP) on local living environment – the attitude of the floodplain residents. Natral Hazard 29(3):309–324
Chowdhury JU, Salehin M (1987) Floods and their processes. In: Proceedings of the international seminar on evolution of scientific system of flood forecasting and warning in the Ganges, Brahmaputra and Meghna River Basins. Bangladesh National Committee of International Commission for Irrigation and Drainage (ICID), Dhaka, pp 247–254
Chowdhury MR, Sato T (2000) The impact of household characteristics on flood damage – a case study. J Floodplain Manage 1(2):35–56
Chowdhury MR, Ward MN (2007) Seasonal flooding in Bangladesh – variability and predictability. Hydrol Process 21(3):335–347
Chowdhury JU, Rahman R, Bala SK, Islam AKMS (1998) Impact of 1998 flood on Dhaka City and performance of flood control works. Institute of Flood Control and Drainage Research, Bangladesh University of Engineering and Technology, Dhaka
Coleman D, Schofield R (eds) (1986) The state of population theory: forward from Malthus. Basil Blackwell, Oxford
Cook BR (2010) Flood knowledge and management in Bangladesh: increasing diversity, complexity and uncertainty. Geogr Compass 4(7):750–767
Cova TJ (1999) GIS in emergency management. In: Longley PA, Goodchild MF, Maguire DJ, Rhind DV (eds) Geographical information systems, management and application. Wiley, New York, pp 845–858
Cross JA (2001) Megacities and small towns: different perspectives on hazard vulnerability. Environ Hazard 3(2):63–80

Cruz RV, Harasawa H, Lal M, Wu S, Anokhin Y, Punsalmaa B, Honda Y, Jafari M, Li C, Ninh NH (2007) Asia. In: Parry ML, Canziani OF, Palutikof JP, van der Linden PJ, Hanson CE (eds) Climate change 2007: impacts adaptation and vulnerability. Contribution of Working Group II to the fourth assessment report of the Intergovernmental Panel on Climate Change. Cambridge University Press, Cambridge, pp 469–506

Custers P (1992) Banking on a flood free future? Flood mismanagement in Bangladesh. The Ecologist 22(5):241–247

Cutter SL (1996) Vulnerability to environmental hazards. Prog Hum Geogr 20(4):529–539

Cutter SL (2003) GI science, disasters, and emergency management. Trans GIS 7(4):439–445

Cutter SL, Finch C (2008) Temporal and spatial changes in social vulnerability to natural hazards. Proc Natl Acad Sci 105(7):2301–2306

Cutter SL, Mitchell JT, Scott MS (2000) Revealing vulnerability of people and places: a case study of Georgetown County, South Carolina. Ann Assoc Am Geogr 90(4):713–737

Cutter SL, Boruff BJ, Shirley WL (2003) Social vulnerability to environmental hazards. Social Sci Q 84(2):242–261

Cutter SL, Emrich CT, Webb JJ, Morath D (2009) Social vulnerability to climate variability hazards: a review of literature. Final report to Oxfam America, available at http://adapt.oxfamamerica.org/resources/Literature_Review.pdf. Accessed 2 Apr, 2011

Dartmouth Flood Observatory (DFO) (2012) Global register of major flood events. http://www.dartmouth.edu/~floods/Currentt.htm. Accessed 10 Feb, 2012

Dasgupta A (2007) Floods and poverty traps: evidence from Bangladesh. Econ Pol Wkly July:3166–3171

Davis I (1987) Safe shelter within unsafe cities. Open House Int 12(3):5–15

del Ninno C, Dorosh PA, Islam N (2002) Reducing vulnerability to natural hazards: lessons from the 1998 floods in Bangladesh. IDS Bull 33(4):98–107

Dewan AM, Yamaguchi Y (2008) Effects of land cover changes on flooding: example from Greater Dhaka of Bangladesh. Int J Geoinf 4(1):11–20

Dewan AM, Nishigaki M, Komatsu M (2003) Floods in Bangladesh: a comparative hydrological investigation on two catastrophic events. J Fac Environ Sci Technol 8(1):53–62

Dewan AM, Yeboah KK, Nishigaki M (2006) Using synthetic aperture radar (SAR) data for mapping river water flooding in an urban landscape: a case study of Greater Dhaka, Bangladesh. J Hydrol Water Resour 19(1):44–54

Dewan AM, Kabir MH, Nahar K, Rahman MZ (2012) Urbanization and environmental degradation in Dhaka metropolitan area of Bangladesh. Int J Environ Sustain Dev 11(2): 118–146

Disaster Management Bureau (DMB) (1998) Report on Bangladesh flood 1998: chronology, damages and responses. DMB, Dhaka

Dodson B (1996) Vulnerability to flooding in rural Bangladesh: a socioeconomic appraisal. In: Singh RB (ed) Disasters, environment and development (Proceedings of International Geographical Union Seminar, New Delhi, 9–12 December 1994). Oxford/IBH Publication, New Delhi/Calcutta, pp 183–195

Du N, Ottens H, Sliuzas R (2010) Spatial impact of urban expansion on surface water bodies – a case study of Wuhan, China. Landscape Urban Plan 94(3–4):175–185

Emch M (2000) Relationships between flood control, kala-azar, and diarrheal disease in Bangladesh. Environ Plan A 32(6):1051–1063

EM-DAT (2012) Global flood records, 1950–2011. Centre for Research on the Epidemiology of Disasters (CRED), Universite catholique de Louvain, Belgium. www.emdat.be. Accessed 15 June, 2012

Etkin D (1999) Risk transference and related trends: driving forces towards more mega-disasters. Environ Hazard 1(2):69–75

Faisal IM, Kabir MR, Nishat A (2003) The disastrous floods of 1998 and long term mitigation strategies for Dhaka City. Nat Hazard 28(1):85–99

Ferdous M, Hossain M (2005) Flood frequency analysis at different rivers in Bangladesh: a comparison study on probability distribution functions. Thammasat Int J Sci Technol 10(3):53–66

Ferreira S, Hamilton K, Vincent JR (2011) Nature, socioeconomics and adaptation to natural hazards: new evidence from floods. Policy research working paper. The World Bank, Washington, DC

Few R (2003) Flooding, vulnerability and coping strategies: local responses to a global threat. Prog Dev 3(1):43–58

Finch C, Emrich CT, Cutter SL (2010) Disaster disparities and differential recovery in New Orleans. Popul Environ 31(4):179–202

Food and Agricultural Organization (FAO) (2004) Gateway to land and water information: Bangladesh national report. http://www.fao.org/ag/agl/swlwpnr/reports/y_sa/z_bd/bd.tm. Accessed 23 Sept, 2004

Gao J, Nickum JE, Pan Y (2007) An assessment of flood hazard vulnerability in the Dongting lake region of China. Lakes Reserv Res Manage 12(1):27–34

Gilbert A (2005) The challenge of slums: global report on human settlements 2003. Prog Hum Geogr 29(1):118–120

Gillis AM (1993) Managing rivers. Bioscience 43(10):677–678

Girard LF, Forte B, Cerreta M, Toro PD, Forte F (eds) (2003) The human sustainable city: challenges and perspective from the habitat agenda. Ashgate Publishing Company, Burlington

Godschalk DR (2003) Urban hazard mitigation: creating resilient cities. Nat Hazard Rev 4(3):136–143

Goudie A (1990) The human impact on the natural environment. The MIT Press, Cambridge

Green CH, Tunstall SM, Fordham M (1991) The risks from flooding: which risk and whose perception. Disasters 15(3):227–236

Greiving S, Fleischhauer M, Luckenkotter J (2006) A methodology for an integrated risk assessment of spatially relevant hazards. J Environ Plan Manage 49(1):1–19

Gupta TN (1994) Vulnerability of houses in hazard prone areas. The world conference on the IDNDR, Yokohama, Japan, 23–27 May

Gupta K (2007) Urban flood resilience planning and management and lessons for the future: a case study of Mumbai, India. Urban Water J 4(3):183–194

Gupta H, Chakrapani GJ (2007) Temporal and spatial variations in water flow and sediment load in the Narmada River. Curr Sci 92(5):679–684

Halls AS, Payne AI, Alam SS, Barman SK (2008) Impacts of flood schemes on inland fisheries in Bangladesh: guidelines for mitigation. Hydrobiologia 609(1):45–58

Hamilton LS (1987) What are the impacts of Himalayan deforestation on the Ganges-Brahmaputra lowlands and delta? Assumptions and facts. Mt Res Dev 7(3):256–263

Hamza M, Zetter R (1998) Structural adjustment, urban systems, and disaster vulnerability in developing countries. Cities 15(4):291–299

Haque CE (1994) Flood prevention and mitigation in Bangladesh: the need for sustainable floodplain development. In: Goodland R, Edmundson V (eds) Environmental impact assessment and development. The World Bank, Washington, DC, pp 101–112

Haque MI (2008) Water resources: management in Bangladesh. Anushilan Publishers, Dhaka

Haque CE, Blair D (1992) Vulnerability to tropical cyclones: evidence from the April cyclone in coastal Bangladesh. Disasters 16(3):217–229

Haque CE, Zaman MQ (1993) Human responses to riverine hazards in Bangladesh: a proposal for sustainable floodplain development. World Dev 21(1):93–107

Haque E, Zaman M (1994) Vulnerability and responses to riverine hazards in Bangladesh: a critique of flood control and mitigation approaches. In: Varley A (ed) Disasters, development and environment. Wiley, London, pp 65–79

Havlick SW (1986) Third world cities at risk: building for calamity. Environment 28(9):6–11, 41–45

Hewitt K (1997) Regions of risk: a geographical introduction to disasters. Longman, Essex

Hirabayashi Y, Kanae S (2009) First estimate of the future global population at risk of flooding. Hydrol Res Lett 3:6–9

Hirabayashi Y, Kanae S, Emori S, Oki T, Kimoto M (2008) Global projections of changing risk of floods and droughts in a changing climate. Hydrol Sci J 53(4):754–773

Hirsch RM, Walker JF, Day JC, Kallio R (1990) The influence of man on hydrological systems. In: Wolman MG, Riggs HC (eds) Surface water hydrology, vol 0-1. Geological Society of America, Boulder, pp 329–359
Hochrainer S, Mechler R (2011) Natural disaster risk in Asian megacities a case for risk pooling? Cities 28(1):53–61
Hofer T, Messerli B (2006) Floods in Bangladesh: history, dynamics and rethinking the role of the Himalayas. United Nations University, Tokyo
Hollis GE (1975) The effects of urbanization on floods of different recurrence interval. Water Resour Res 11(3):431–435
Hoque MM, Siddique AB (1995) Flood control projects in Bangladesh: reasons for failure and recommendations for improvement. Disasters 19(3):260–263
Hoque R, Nakayama D, Matsuyama H, Matsumoto J (2011) Flood monitoring, mapping and assessing capabilities using RADARSAT remote sensing, GIS and ground data for Bangladesh. Nat Hazard 57(2):525–548
Horlick-Jones T (1995) Urban disasters and megacities in a risk society. GeoJournal 37(3):329–334
Hossain MM (1998) Bangladesh floods and its management initiatives. Steps Towards Dev 3:12–19
Hossain ANH (2004) The impact of floods on Bangladesh and options for mitigation an overview. In: Siddiqui KU, Hossain ANH (eds) Options for flood risk and damage reduction in Bangladesh. The University Press Limited, Dhaka, pp 55–70
Hughes R, Adnan S, Dalal-Clayton B (1994) Flood plains and flood plans? A review of approaches to water management in Bangladesh. Research and Advisory Services, Dhaka
Hunt A, Watkiss P (2011) Climate change impacts and adaptation in cities: a review of the literature. Clim Chang 104(1):13–49
Huq S (1999) Environmental hazards in Dhaka. In: Mitchell JK (ed) Crucibles of hazards: megacities and disasters in transition. United Nations University, Tokyo, pp 119–137
Hutton D, Haque CE (2004) Human vulnerability, dislocation and resettlement: adaptation process of river bank erosion-induced displaces in Bangladesh. Disasters 28(1):41–62
International Federation of Red Cross and Red Crescent Societies (IFRC/RCS) (2010) World disasters report 2010: focus on urban risk. IFRC/RCS, Geneva
IPCC (2007a) Impacts, adaptation and vulnerability. Cambridge University Press, Cambridge
IPCC (2007b) Climate change 2007: synthesis report. Contributions of Working Groups I, II, III to the fourth assessment report of the Intergovernmental Panel on Climate Change. IPCC, Geneva
Irfanullah HM, Azad MAK, Kamruzzaman M, Wahed MA (2011) Floating gardening in Bangladesh: a means to rebuild lives after devastating flood. Indian J Tradit Knowl 10(1):31–38
Islam MZ (1991) Failure of flood embankments: case studies of some selected projects in Bangladesh. Final report, R02/91. Institute of Flood Control and Drainage Research (IFCDR), BUET, Dhaka
Islam N (2001) The open approach to flood control: the way to the future in Bangladesh. Futures 33(8–9):783–802
Islam KMN (2006) Impacts of flood in urban Bangladesh: micro and macro level analysis. A H Development Publishing House/Community Development Library, Dhaka
Islam MM, Sado K (2000a) Satellite remote sensing data analysis for flood damage zoning with GIS for flood management. Annu J Hydraulic Eng (JSCE) 44:301–306
Islam MM, Sado K (2000b) Development of flood hazard maps of Bangladesh using NOAA AVHRR with GIS. Hydrol Sci J 45(3):337–355
Islam MM, Sado K (2000c) Flood hazard assessment in Bangladesh using NOAA AVHRR data with geographical information system. Hydrol Process 14(3):605–620
Islam AKMS, Haque A, Bala SK (2010) Hydrologic characteristics of floods in Ganges-Brahmaputra-Meghna (GBM) delta. Nat Hazard 54(3):797–811
Ives J (1991) Floods in Bangladesh – who is to blame. New Sci 130(1764):34–37
Jabeen H, Johnson C, Allen A (2010) Built-in resilience: learning from grassroots coping strategies for climate variability. Environ Urban 22(2):415–431

Jacobs JW, Wescoat JL Jr (1994) Flood-hazard problems and programs in Asia's large river basins. Asian J Environ Manage 2(2):91–104
Jakobsen F, Hoque AKMZ, Paudyal GN, Bhuiyan MS (2005) Evaluation of the short-term forcing the monsoon river floods in Bangladesh. Water Int 30(3):389–399
Jha AK, Bloch R, Lamond J (2012) Cities and flooding: a guide to integrated urban flood risk management for the 21st century. The World Bank, Washington, DC
Johnson C (2010) Urban disaster trends. In: World disasters report – focus on urban risk. International Federation of Red Cross and Red Crescent Societies, Geneva, pp 31–51
Jones BG, Kandel WA (1992) Population growth, urbanization, disaster risk, and vulnerability in metropolitan areas: a conceptual framework. In: Kreimer A, Munasinghe M (eds) Environmental management and urban vulnerability. The World Bank, Washington, DC, pp 51–76
Jonkman SN (2005) Global perspectives on loss of human lives caused by floods. Nat Hazard 34(2):151–175
Kale V (2003) Geomorphic effect of monsoon floods on Indian rivers. Nat Hazard 28(1):65–84
Kale V (2012) On the link between extreme floods and excess monsoon epochs in South Asia. Clim Dyn 39(5):1107–1122. doi:10.1007/s00382-011-1251-6
Karim N (1995) Disasters in Bangladesh. Nat Hazards 11(3):247–258
Kasperson RE (1986) Six propositions on public participation and their relevance for risk communication. Risk Anal 6(3):275–281
Kates RW (1996) Human adjustment. In: Hanson S (ed) Ten geographic ideas that changed the world. Rutgers University Press, New Brunswick, pp 87–107
Keller EA, Blodgett RH (2008) Natural hazards: earth's processes as hazards, disasters and catastrophes, 2nd edn. Prentice Hall, Upper Saddle River
Kelly C (1995) Assessing disaster needs in megacities, perspectives from developing countries. GeoJournal 37(3):381–385
Khalequzzaman M (1994) Recent floods in Bangladesh: possible causes and solutions. Nat Hazard 9(1–2):65–80
Khalil MG (1990) Floods in Bangladesh: a question of disciplining the rivers. Nat Hazard 3(4):379–401
Khalil MG, Hossain MM, Hoque MM (1995) On siltation in the rivers of Bangladesh: causes and consequences. J Natl Oceanogr Marit Inst (NOAMI) 12(1 & 2):11–20
Khan A (1999) The 1998 Bangladesh floods and flood security. In: Harvard University Asia Centre (ed) Natural disasters and policy response Asia: implications for food security. Harvard University Asia Centre, Cambridge, MA
Khan MSA (2008) Disaster preparedness for sustainable development in Bangladesh. Disast Prevt Manage 17(5):662–671
Kibler DF, Froelich CD, Aron G (1981) Analyzing urbanization impacts on Pennsylvania flood peaks. J Am Water Resour Assoc 17(2):270–274
Kibria AMMG (1970) Protection of East Pakistan against floods and cyclonic surges. Pak Eng 10(1):529–572
Kienberger S (2012) Spatial modelling of social and economic vulnerability to floods at the district level of Buzi, Mozambique. Nat Hazard. doi:10.1007/s11069-012-0174-9
Kleinen T, Petschel-Held G (2007) Integrated assessment of change in flooding probabilities due to climate change. Clim Chang 81(3):283–312
Kraas F (2007) Megacities and global change: key priorities. Geogr J 173(1):79–82
Kraas F (2008) Megacities as global risk areas. In: Marzluff J, Shulenberger E, Endlicher W, Alberti M, Bradley G, Ryan C, ZumBrunnen C, Simon U (eds) Urban ecology: an international perspective on the interaction between humans and nature. Section V. Springer, New York, pp 583–596
Kreimer A, Munasinghe M (1991) The environment and disaster management. Land Use Policy 8(4):269–281
Kubal C, Haase D, Meyer V, Scheuer S (2009) Integrated urban flood risk assessment – adapting a multi-criteria approach to a city. Nat Hazard Earth Syst Sci 9(6):1881–1895

Kubo S (1993) Geomorphological features of northwestern Bangladesh and some problems on flood mitigation. GeoJournal 31(4):313–318
Kumar V, Jain SK, Singh Y (2010) Analysis of long-term rainfall trends in India. Hydrol Sci 55(4):484–496
Kundzewicz ZW, Graczyk D, Maurer T, Pinskwar I, Radziejewski M, Svensson C, Szwed M (2005) Trend detection in river flow series 1: annual maximum flow. Hydrol Sci J 50(5):797–810
Kundzewicz ZW, Hirabayashi Y, Kanae S (2010) River floods in the changing climate – observations and projections. Water Resour Manage 24(11):2633–2646
Kunii O, Nakamura S, Abdur R, Wakai S (2002) The impact on health and risk factors of the diarrhoea epidemics in the 1998 Bangladesh floods. Public Health 116(2):68–74
Lall SV, Deichmann U (2012) Density and disasters: economics of urban hazard risk. World Bank Res Obs 27(1):74–105
Lankao PR, Qin H (2011) Conceptualizing urban vulnerability to global climate and environmental change. Curr Option Environ Sustain 3(3):142–149
Leopold LB (1994) A view of the river. Harvard University Press, Cambridge, MA
Liong SY, Sivapragasam C (2002) Flood stage forecasting with support vector machines. J Am Water Resour Assoc 38(1):173–186
Liong SY, Lim WH, Kojiri T, Hori T (2000) Advance flood forecasting for flood stricken Bangladesh with a fuzzy reasoning method. Hydrol Process 14(3):431–448
Liverman DM (1986) The vulnerability of urban areas to technological risks: an overview of US and European experience. Cities 3(2):142–147
Mahmud W (2004) The impact of floods on the economy of Bangladesh. In: Siddiqui KU, Hossain ANH (eds) Options for flood risk and damage reduction in Bangladesh. The University Press Limited, Dhaka, pp 81–92
Menoni S, Pergalani F (1996) An attempt to link risk assessment with land use planning: a recent experience in Italy. Disast Prev Manage 5(1):6–21
Meyer V, Scheuer S, Haase D (2009) A multicriteria approach for flood risk mapping exemplified at the Mulde river, Germany. Natura Hazard 48(1):17–39
Milly PCD, Wetherald RT, Dunne KA, Delworth TL (2002) Increasing risk of great floods in a changing climate. Nature 415:514–517
Milly PCD, Betancourt J, Falkenmark M, Hirsch RM, Kundzewicz ZW, Lettenmaier DP, Stouffer RJ (2008) Stationarity is dead: whither water management? Science 319:573–574
Minkin SF, Rahman R, Islam MA (1996) Flood control embankments and epidemic kala-azar in Bangladesh. Ecosyst Health 2:215–226
Mirza MMQ (1984) Flood has become a nightmare. Bangladesh Today 2:26–35
Mirza MMQ (1991) Flood action plan of Bangladesh – the embankment issue. Water Nepal 2(2/3):25–28
Mirza MMQ (1997) Modelling the effects of climate change on flooding in Bangladesh. PhD thesis, IGCI, University of Waikato, New Zealand
Mirza MMQ (2002) Global warming and changes in the probability of occurrence of floods in Bangladesh and implications. Glob Environ Chang 12(2):127–138
Mirza MMQ (2003) Three recent extreme floods in Bangladesh: a hydro-meteorological analysis. Nat Hazard 28(1):35–64
Mirza MMQ (2011) Climate change, flooding in South Asia and implications. Reg Environ Chang 1(supp):95–107
Mirza MMQ, Ericksen NJ (1996) Impact of water control projects on fisheries resources in Bangladesh. Environ Manage 20(4):523–539
Mirza MMQ, Warrick RA, Ericksen NJ (2003) The implications of climate change on floods of the Ganges, Brahmaputra and Meghna Rivers in Bangladesh. Clim Chang 57(3):287–318
Mitchell JK (1989) Hazard research. In: Gaile GL, Willmott CJ (eds) Geography in America. Morril, Columbus, pp 410–424
Mitchell JK (1993) Natural hazard predictions and responses in very large cities. In: Nemec J, Nigg JM, Siccardi F (eds) Predictions and perceptions of natural hazards. Kluwer Academic Publishers, Dordrecht, pp 29–37

Mitchell JK (1999a) Natural disasters in the context of mega-cities. In: Mitchell JK (ed) Crucibles of hazard: mega-cities and disasters in transition. United Nations University, Tokyo, pp 15–55
Mitchell JK (1999b) Megacities and natural disasters: a comparative analysis. GeoJournal 49(2):137–142
Mitchell JK, Devine N, Jagger K (1989) A contextual model of natural hazard. Geogr Rev 79(4):391–409
Montoya A (2003) Geo-data acquisition through mobile GIS and digital video: an urban disaster management perspective. Environ Model Software 18(10):869–876
Montz BE, Tobin GA (2003) Hazardousness of the Tampa region: evaluating physical risk and socio-economic vulnerability. Proc Appl Geogr Conf 31:380–388
Moore J (2001) Cities at risk. Habitat Debate 7(4):1–6
Morrow BH (1999) Identifying and mapping community vulnerability. Disasters 23(1):1–18
Müller A (2012) Areas at risk – concept and methods for urban flood risk assessment: a case study of Santiago de Chile. Franz Steiner Verlag, Stuttgart
Munich Re (2004) Megacities – megarisks: trends and challenges for insurance and risk management. Münchener Rückversicherungs-Gesellschaft, Munchen
Myers MF, White GF (1993) The challenge of the Mississippi flood. Environment 35(10):6–25
Nakashima S, Khan MH (1994) A basic guide to understanding the environmental impacts of rural roads on the wetlands of Bangladesh. Care International Bangladesh, Dhaka
Nchito WS (2007) Flood risk in unplanned settlements in Lusaka. Environ Urban 19(2):539–551
Nicholls RJ (1995) Coastal megacities and climate change. GeoJournal 37(3):369–379
Nirupama N, Simonovic SP (2007) Increase of flood risk due to urbanization: a Canadian example. Natura Haz 40(1):25–41
O'Brien K, Leichenko R, Kelkar V, Venema H, Aandahl G, Tompkins H, Javed A, Bhadwal S, Barg S, Nygaard L, West J (2004) Mapping vulnerability to multiple stressors: climate change and globalization in India. Glob Environ Chang 14(4):303–313
Odeh DJ (2002) Natural hazards vulnerability assessment for statewide mitigation planning in Rhode island. Nat Hazard Rev 3(4):177–187
Osti R, Hishinuma S, Miyake K, Inomata H (2011) Lessons learned from statistical comparison of flood impact factors among southern and eastern Asian countries. J Flood Risk Manage 4(3):203–215
Oya M (1990) Relationship between geomorphology and flooding in the Brahmaputra-Jamuna and Ganges floodplain based on the geomorphological land classification map indicating areas subject to flooding and utilization for the mitigation of flood damages. Nat Disast Sci 9(2):1–17
Pall P, Aina T, Stone DA, Stott PA, Nozawa T, Hilberts AGJ, Lohmann D, Allen MR (2011) Anthropogenic greenhouse gas contribution to flood risk in England and Wales in autumn 2000. Nature 470:382–386
Palmer TN, Rälsänen J (2002) Quantifying the risk of seasonal precipitation events in a changing climate. Nature 415:512–514
Parker D (1999) Flood. In: Ingleton J (ed) Natural disaster management. Tudor Rose, Leicester, pp 38–40
Parnell S, Simon D, Vogel C (2007) Global environmental change: conceptualising the growing challenge for cities in poor countries. Area 39(3):357–369
Paudyal GN (2002) Forecasting and warning of water-related disasters in a complex hydraulic setting-the case of Bangladesh. Hydrol Sci J 47:S5–S18
Paul BK (1984) Perception and agricultural adjustments to floods in Jamuna floodplain, Bangladesh. Hum Ecol 12(1):3–19
Paul BK (1997) Flood research in Bangladesh in retrospect and prospect: a review. Geoforum 28:121–131
Paul MJ, Meyer JL (2001) Streams in urban landscape. Annu Rev Ecol Syst 32:333–365
Pelling M (2002) Assessing urban vulnerability and social adaptation to risk. Int Dev Plan Rev 24(1):59–76
Pelling M (2003) The vulnerability of cities. Earthscan, London

Quarantelli EL (2003) Urban vulnerability to disasters in developing countries: managing risks. In: Kreimer A, Arnold M, Carlin A (eds) Building safer cities: the futures of disaster risk. The World Bank, Washington, DC, pp 211–232

Rabbani MG (2009) Environmental risks in Dhaka: present initiatives and future improvements. In: Shaw R, Srinivas H, Sharma A (eds) Urban risk reduction: an Asian perspective. Emerald Group Publishing, Bingley, pp 319–338

Rahman R, Haque A, Khan SA, Salehin M, Bala SK (2005) Investigation of hydrological aspects of flood – 2004 with special emphasis on Dhaka City. Final report. Institute of Water and Flood Management, BUET, Dhaka

Rana MY, Lambert MF (2000) Generating a flood level probability map for Bangladesh. Rural Environ Eng 38(2):30–39

Ranger N, Hallegatte S, Bhattacharya S, Bachu M, Priya S, Dhore K, Rafique F, Mathur P, Naville N, Henriet F, Herweijer C, Pohit S, Corfee-Morlot J (2011) An assessment of the potential impact of climate change on flood risk in Mumbai. Clim Chang 104:139–167

Rashed T, Weeks J (2003) Assessing vulnerability to earthquake hazards through spatial multicriteria analysis of urban areas. Int J Geogr Inf Sci 17(6):547–576

Rashed R, Weeks J, Couclelis H, Herold M (2007) An integrative GIS and remote sensing model for place-based urban vulnerability analysis. In: Mesev V (ed) Integration of GIS and remote sensing. Wiley, Chichester, pp 199–224

Rasheed KBS (2008) Bangladesh: resource and environmental profile. A H Development Publishing House, Dhaka

Rashid S (1991) Flood action plan: a view from abroad. Grassroots 1(1):8–11

Rashid SF (2000) The urban poor in Dhaka city: their struggles and coping strategies during the floods of 1998. Disasters 24(3):240–253

Rasid H, Mallik A (1995) Flood adaptations in Bangladesh: is the compartmentalization scheme compatible with indigenous adjustments of rice cropping to flood regimes? Appl Geogr 15(1):3–17

Rasid H, Mallik A (1996) Living on the edge of stagnant water: an assessment of environmental impacts of construction-phase drainage congestion along Dhaka City flood control embankment, Bangladesh. Environ Manage 20(1):89–98

Rasid H, Paul BK (1987) Flood problems in Bangladesh: is there an indigenous solution? Environ Manage 11(2):155–173

Rayhan MI (2010) Assessing poverty, risk and vulnerability: a study on flooded households in rural Bangladesh. J Flood Risk Management 3(1):18–24

Reavill LRP, Rahman TG (1995) A systems-science-based analysis of the factors that influence and aggravate the effects of flooding in Bangladesh. Technol Forecast Soc Chang 49(1):89–101

Rob MA (1990) Flood hazard in Bangladesh: nature, causes and control. Asian Profile 18(4):365–378

Rogers P, Lydon P, Secklet D (1989) Eastern water study: strategies to manage flood and drought in the Ganges-Brahmaputra basin. Irrigation Support Project for Asia and the Near East (ISPAN), Arlington

Rosenfield C (1994) Flood hazard reduction: GIS maps survival strategies in Bangladesh. Geo Inf Syst 4(5):29–37

Saleh AFM, Mondal MS (1999) Performance of flood control and drainage projects during the 1998 flood: a case study. J Indian Water Resour Soc 19(4):1–8

Sanderson D (2000) Cities, disasters and livelihoods. Environ Urban 12(2):93–102

Sanderson D (2012) Building livelihoods to reduce risk among the most marginalized in urban areas: strategic approaches from Dhaka. Environ Hazard 11(2):112–122

Sanyal J, Lu XX (2004) Application of remote sensing in flood management with special reference to Monsoon Asia: a review. Nat Hazard 33(2):283–301

Sanyal J, Lu XX (2009) Ideal location for flood shelter: a geographic information system approach. J Flood Risk Assess 2(4):262–271

Satterthwaite D (2010) Avoiding the urbanization of disasters. In: World disasters report – focus on urban risk. International Federation of Red Cross and Red Crescent Societies, Geneva, pp 11–29

Schanze J (2006) Flood risk management – a basic framework. In: Schanze J, Zeman E, Marsalek J (eds) Flood risk management – hazards, vulnerability and mitigation measures. Springer, Ostrov, pp 149–167

Senga R (2004) Natural and unnatural disasters: the relative vulnerabilities of Southeast Asian megacities to climate change. World Wide Fund for Nature (WWF), Switzerland. www2.lse.ac.uk/IDEAS/publications/reports/pdf/SR004/WWF.pdf. Accessed 15 Mar, 2012

Seto KC, Sánchez-Rodríguez R, Fragkias M (2010) The new geography of contemporary urbanization and the environment. Annu Rev Environ Resour 35:167–194

Shailo I (1988) Glimpses into the history of the floods: 35 Years. ADAB News Sept–Oct:7–8

Sharma A, Surjan A, Shaw R (2011) Overview of urban development and associated risks. In: Sharma A, Surjan A, Shaw R (eds) Community, environment and disaster risk management. Emerald Group Publishing, Bingley, pp 1–16

Sherbinin AD, Schiller A, Pulsipher A (2007) The vulnerability of global cities to climate hazards. Environ Urban 19(1):39–64

Shimi AC, Parvin GA, Biswas C, Shaw R (2010) Impact and adaptation to flood: a focus on water sully, sanitation and health problems of rural community in Bangladesh. Disast Prev Manage 19(3):298–313

Showalter PS, Lu Y (eds) (2010) Geospatial techniques in urban hazards and disaster analysis. Geotechnologies and the environment 2. Springer, Dordrecht

Siddiqui KU, Hossain ANHA (2004) Options for flood risk and damage reduction in Bangladesh. The University Press Limited, Dhaka

Sklar L (1993) Drowning in aid: the World's Bank Bangladesh Flood Action Plan. Multinatl Monit 14(4):8–13

Smith K (2001) Environmental hazards: assessing risk and reducing disaster, 3rd edn. Routledge, London

Solomon S, Qin D, Manning M, Chen Z, Marquis M, Averyt KB, Tignor M, Miller HL (eds) (2007) Climate change 2007: the physical science basis. Contribution to Working Group I to the fourth assessment report of Intergovernmental Panel on Climate Change. Cambridge University Press, Cambridge

Solway L (1994) Urban developments and megacities: vulnerability to natural disasters. Disast Manage 6(3):160–169

Solway L (1999) Socio-economic perspective of developing country megacities vulnerable to flood and landslide hazards. In: Casale R, Margottini C (eds) Floods and landslides: integrated risk assessment. Springer, Berlin, pp 245–260

Srinivas H, Shaw R, Sharma A (2009) Introduction to urban risk reduction. In: Shaw R, Srinivas H, Sharma A (eds) Urban risk reduction: an Asian perspective. Emerald Group Publishing, Bingley, pp 3–12

Stalenberg B, Vrijling H (2009) The battle of Tokyo and Dhaka against floods. Built Environ 35(4):471–491

Steedman S (1995) Megacities: the unacceptable risk natural disaster. Built Environ 21(2/3): 89–94

Stewart K (1988) Post flood: assessment and nutritional status of children in Matlab, Bangladesh. Paper presented at the seminar on regional and global environmental perspectives, Dhaka, 4–6 March

Sultana P, Thompson PM (1997) Effects of flood control and drainage on fisheries in Bangladesh and the design of mitigating measures. Regul Rivers: Res Manage 13(1):43–55

Suriya S, Mudgal BV (2012) Impact of urbanization on flooding: the Thirusoolam sub watershed – a case study. J Hydrol 412–413:210–219

Szlafsztein C, Sterr H (2007) A GIS-based vulnerability assessment of coastal natural hazards, state of Para, Brazil. J Coast Conserv 11(1):53–66

Tapsell SM, Penning-Rowsell C, Tunstall SM, Wilson TL (2002) Vulnerability to flooding: health and social dimensions. Philos Trans R Soc Lond A 360(1796):1511–1525

Tarbuck EJ, Lutgens FK, Tasa D (2005) Earth: an introduction to physical geology, 8th edn. Pearson Education Inc, Upper Saddle River

Taubenböck H, Post J, Roth A, Zosseder K, Strunz G, Dech S (2008) A conceptual vulnerability and risk framework as outline to identify capabilities of remote sensing. Nat Hazard Earth Syst Sci 8(3):409–420
Taubenböck H, Roth A, Dech S (2009) Megacities: hints for risk assessment using EO data. In: Gamba P, Herold M (eds) Global mapping of human settlement: experiences, datasets and prospects. CRC Press, Boca Raton, pp 205–230
Thomas DSK, Mitchell JT (2001) Which are the most hazardous states? In: Cutter SL (ed) American hazardscapes: the regionalization of hazards and disasters. Joseph Henry Press, Washington, DC, pp 115–156
Thompson PM (1996) Operation and maintenance performance and conflicts in flood-control projects in Bangladesh. Int J Water Resour Dev 12(3):311–328
Tipple G (2006) Housing, urban vulnerability and sustainability in rapidly – developing cities. Built Environ 32(4):387–399
Tobin GA, Montz BE (1997) Natural hazards: explanation and integration. Guilford Press, New York
Tol RSJ (2008) Why worry about climate change? A research agenda. Environ Value 17(4):437–470
Tran P, Kaneko F, Shaw R, Victoria LP, Oi H (2009a) Urban disaster risk analysis, action planning and implementation management. In: Shaw R, Srinivas H, Sharma A (eds) Urban risk reduction: an Asian perspective. Emerald Group Publishing, Bingley, pp 13–36
Tran P, Shaw R, Chantry G, Norton J (2009b) GIS and local knowledge in disaster management: a case study of flood risk mapping in Vietnam. Disasters 33(1):152–169
Uitto JI (1998) The geography of disaster vulnerability in megacities. Appl Geogr 18(1):7–16
Ulusay R, Aydan O, Kumsar H, Sonmez H (2002) Engineering geological characteristics of the 1998 Adana-Ceyhan earthquake, with particular emphasis on liquefaction phenomena and the role of soil behaviour. Bull Eng Geol Environ 59(2):99–118
United Nations (UN) (2003) Guidelines for reducing flood losses. Available at http://www.un.org/esa/sustdev/publications/flood_guidelines.pdf. Accessed 12 Jan, 2010
United Nations (2012) World population prospects: the 2010 revision, vol 1. United Nations Department of Economic and Social Affairs, New York. Available at: http://eas.un.org/undp/wpp/Documentation/publications.html. Accessed 1 Jan, 2012
United Nations Human Settlements Program (UNHSP) (2004) Reducing urban risk and vulnerability: a thematic paper submitted for discussion at the UN-HABITAT/UN-ISDR. Working meeting on vulnerability assessment and reducing urban risk, Madrid, 7–9 September
United Nations Human Settlements Programme (UN-HABITAT) (2010) State of the World's Cities 2010/2011 – bridging the urban divide. UN-HABITAT, Washington, DC
United Nations International Strategy for Disaster Reduction (UNISDR) (2005) Hyogo framework for action 2005–2015: building the resilience of nations and communities to disasters. United Nations, Geneva
van Aalast MK (2006) The impact of climate change on the risk of natural disasters. Disasters 30(1):5–18
Varis O, Kummu M, Salmivaara A (2011) Ten major rivers in monsoon Asia-Pacific: an assessment of vulnerability. Appl Geogr 32(2):441–451
Water Resources Planning Organization (WARPO) (2000) National water management plan: draft development strategy, volume 2, main report. WARPO, Ministry of Water Resources, Dhaka
Weichselgartner J (2001) Disaster mitigation: the concept of vulnerability revisited. Disast Prev Manage 10(2):85–94
Weng Q (2001) Modeling urban growth effects on surface runoff with the integration of remote sensing and GIS. Environ Manage 28(6):737–748
Wenzel F, Bendimerad F, Sinha R (2007) Megacities – megarisks. Nat Hazard 42(3):481–491
Wetherald RT, Manabe S (2002) Simulation of hydrologic changes associated with global warming. J Geophys Res 107(4379):15
White GF (ed) (1974) Natural hazards: local, national, global. Oxford University Press, New York
White GF, Kates RW, Burton I (2001) Knowing better and losing even more: the use of knowledge in hazards management. Environ Hazard 3(3–4):81–92

Whitfield PH (2012) Floods in future climates: a review. J Flood Risk Manag 5(4):336–365
Wilby RL, Keenan R (2012) Adapting to flood risk under climate change. Prog Phys Geogr 36(3):348–378
Wisner B (2003) Disaster risk reduction in megacities: making the most of human and social capital. In: Kreimer A, Arnold M, Carlin A (eds) Building safer cities: the futures of disaster risk. The World Bank, Washington, DC, pp 181–196
World Bank (2011a) Urban risk assessments: an approach for understanding disaster and climate risk in cities. Urban Development and Local Government Unit, Washington, DC
World Bank (2011b). The World Bank supports Thailand's post-floods recovery effort. The World Bank. http://go.worldbank.org/TCFEHXJML0. Accessed 20 Feb, 2012
World Bank, UN-ISDR (2008) Climate resilient cities: a primer on reducing vulnerabilities to climate change impacts and strengthening disaster risk management in East Asian cities. The World Bank, Washington, DC
Yoon DK (2012) Assessment of social vulnerability to natural disasters: a comparative study. Nat Hazard 63(2):823–843
Zahran S, Brody SD, Peacock WG, Vedlitz A, Gorver H (2008) Social vulnerability and the natural and built environment: a model of flood causalities in Texas. Disasters 32(4):537–560
Zahurul M (1991) Embankment failure in Bangladesh: causes and recommendations. Grassroots 1(2):20–41

Chapter 2
Hazards, Risk, and Vulnerability

Abstract Natural hazard terminologies with existing models of vulnerability, hazard, and risk are presented in this chapter. A conceptual framework has been developed based on hazard literature. The basic premise of the framework is based on Cutter's place of hazard theory with inputs from recent literature. In the framework, hazard is viewed as a threat that has the potential to overwhelm people, property, and the environment. It is a pre-existing condition that can turn into a catastrophe depending on the influence of exogenous and endogenous factors. Exposure to hazard is treated as given and is an implicit element. The vulnerability element is perceived as the interactive effects of the social and physical aspects of a system (e.g., urban) regarding the causal process of hazards. Contrary to some conceptualizations, the framework views that the total vulnerability of a community depends on physical, social, and existing coping capacity attributes, and therefore, the calculation of the total vulnerability should consider these elements simultaneously. Risk is conceptualized as the product of hazard and vulnerability. To minimize the effects of disasters, it is imperative to take appropriate measures to reduce vulnerability rather than risk.

The utilization of geospatial techniques in flood risk management is separated into three categories: flood mapping, damage assessment, and evaluation of flood risk and vulnerability. Biophysical and socioeconomic data that are sourced from remote sensing, census geography, and other spatial databases are employed to evaluate flood-related risk in diverse environments. A variety of methods—comprising inductive, deductive, and multi-criteria evaluation—are used to determine flood vulnerability and risk as evidenced by the literature survey. It reveals that geospatial techniques can be utilized effectively in the entire spectrum of the disaster cycle, which can save lives and property from natural hazards such as flood, as well as support informed decision making during emergencies.

A.M. Dewan, *Floods in a Megacity: Geospatial Techniques in Assessing Hazards, Risk and Vulnerability*, Springer Geography, DOI 10.1007/978-94-007-5875-9_2,

2.1 Introduction

Recent natural hazard literature indicates that there have been several changes in the body of knowledge. For instance, much attention is now being directed towards the understanding of human vulnerability to natural hazards along with other changes (White et al. 2001). As Parker (2000) noted, two factors may have contributed to these changes. Firstly, natural hazard interpretation was viewed as *agent-specific* during much of the twentieth century, and hazards were described as natural phenomena or a threat to the society linked with human modification of the environment. Secondly, a new approach called *social agent* emerged to interpret a natural hazard that underpins the concept of vulnerability. The basic premise of the social agent approach is that disaster is the interaction of hazards and people's vulnerability (Cannon 2000) produced by society (Parker 2000).

As there are several approaches to evaluate the societal risk of natural hazards, the different notions of natural hazards are blurred, making it increasingly difficult to separate natural hazards from technological hazards or social hazards such as violence and war (Pelling 2003; Mitchell 1999a, b; Burton et al. 1993). With the increasing focus of an interdisciplinary flavor in the natural hazard domain, demarcation between the terms may be somewhat difficult when a particular hazard research entails two disciplines, in which case both are multidisciplinary in nature. New approaches have evolved with the advancement of technology, and they have provided ample opportunities to integrate environmental, social, economic, and other parameters to describe societal vulnerability to natural hazards (see Taubenböck et al. 2009). As a result, multiple meanings of key terminologies have mainly originated from epistemological orientations and methodological practices (Mustafa 2005; Cutter 1996).

To understand the different notions associated with natural hazard literature, a number of terminologies are in use to describe hazard, vulnerability, and risk. The definition of these terminologies and a review of the existing models of hazards and vulnerability could be useful in constructing a conceptual framework for this study. In this chapter, Sect. 2.1 presents a brief overview of the concepts of hazard, risk, and vulnerability. Section 2.2 discusses existing models of hazards and vulnerability. Based on an extensive literature review, a conceptual framework has been developed and illustrated in Sect. 2.3. An account of the utilization of geospatial techniques in the assessment of flood risk is described in Sect. 2.4.

2.2 Concept of Vulnerability, Hazard, and Risk

Originally developed in poverty research (Müller 2012), the notion of vulnerability has now become a cornerstone of natural hazard study. Although the concept has been used in geography since the 1970s to describe human–nature interaction (White 1974), it received ample attention after Timmerman's conceptualization in hazard research in the 1980s (Timmerman 1981). Since then, there have been considerable efforts to understand, characterize, and map vulnerability. Literature suggest that there are several lineages on the theoretical construct of vulnerability, which may be divided into three major premises (Müller 2012; Lankao and Qin 2011; Cutter et al. 2008, 2009; Taubenböck et al. 2008; O'Brien et al. 2007; Polsky et al. 2007; Adger 2006; Eakin and Luers 2006; Birkmann 2006; Greiving et al. 2006; Mustafa 2005; Hogan and Marandola 2005; Wisner et al. 2004; UN/ISDR 2004; Oliver-Smith 2004; Bankoff et al. 2004; Pelling 2003; Turner et al. 2003; Bankoff 2001; Mitchell 1999a; Lewis 1999; Hewitt 1997; Dow and Downing 1995; Watts and Bohle 1993; Burton et al. 1993; Dow 1992; Liverman 1990a, b; White and Haas 1975; Timmerman 1981). The first of these stems from the risk and hazard paradigm, which is based on human–nature interaction. This theoretical construct usually answers questions relating to the nature of the hazards, the people that live in hazardous places, and the probable effects. Here, vulnerability is viewed as an outcome of the hazard and is determined by exposure, sensitivity, and potential consequences. The second lineage emphasizes the social dimension of hazards and is rooted in the notion of political economy and political ecology. This theme broadly examines power structure, distribution of resources, and the cultural and economic aspects of a community. This establishes the area that is highly vulnerable, how and why a particular community is vulnerable, and why the effects of hazards on a community or individuals are uneven. The third lineage utilizes the concept of resilience science to comprehend societal vulnerability to global and climate change. As noted by Eakin and Luers (2006, p. 371), '*in this paradigm vulnerability is seen as a dynamic property of a system in which humans are constantly interacting with the biophysical environment.*' Using socioecological system properties, this interdisciplinary and integrative approach aims to develop ideas that can support the goals of sustainability (Adger 2006).

While the concept of vulnerability has extensively been used in hazard research and other disciplines, it lacks an acceptable definition (Cutter et al. 2009; Adger 2006; Green 2004; Weichselgartner 2001; Cutter 1996; Wisner 1993). The conceptualization of vulnerability varies with the topic, discipline, organization, and/or researcher, reflecting different ideological and disciplinary perspectives (Birkmann 2006; UN 2003; Parker 2000; Cutter 1996; Dow 1992). Although there is no intention to review existing definitions related to vulnerability, a comparative evaluation of few definitions may facilitate to recognize the differences. For instance, Dow (1992) defined vulnerability to hazards as '*people's differential incapacity to deal with hazards, based on the position of groups and individuals within both the physical and social worlds.*' Cutter (1996) defined vulnerability to environmental hazard as a '*potential for loss.*' The United Nations Environment Programme (UNEP) (2003) defined vulnerability as the '*manifestation of social, economical, political structure and environmental settings.*' Odeh (2002) referred vulnerability to as '*the combined effect of hazards and exposures in a given region.*' Due to its multifaceted nature, indicators that are used to characterize vulnerability to environmental hazards depend on various factors such as the perspective of research and scale of the study (Fekete et al. 2010; Adger 2006; Green 2004). A wealth of literature exists on the concept of vulnerability and its development (Cutter et al. 2009; Birkmann 2006).

The causal structure of vulnerability also varies significantly. For instance, Watts and Bohle (1993) explain that the social space of vulnerability is the intersection of a tripartite: entitlement, empowerment, and political economy. Blaikie et al. (1994) and Wisner et al. (2004) illustrate the causation of vulnerability with two models, namely, the pressure and release model (PAR) and the access model. The full length of these models is described below. Cutter (1996) identifies three different tracts on the causal structure of vulnerability to environmental hazards. They are: (1) the human ecological perspective, which focuses on the source or potential exposure of biophysical or technological hazards; (2) political ecological perspective, in which vulnerability is seen as a social construct and is rooted in community or individuals' socioeconomic backgrounds; and (3) an integrative approach that focuses on both potential exposure and societal response capability. Spatial context, or location, is the focal point of the latter theme. According to Cutter et al. (2000), the integrative theme is instrumental in dealing with naturally occurring phenomena such as flood and earthquake. Pelling (2003) argued that the vulnerability of a city to environmental hazards depends on three major factors: exposure (location relative to hazard), resistance (livelihood), and resilience (adjustments, preparation). Parallel with causal structure, the measurement of vulnerability also varies, as a variety of techniques and indicators are used at different scales of study (Cutter et al. 2009; Birkmann 2007; Adger 2006; Eakin and Luers 2006).

Vulnerability can be categorized into a number of types, including economical, social, and political, depending on the area being explored (Jha et al. 2012). However, based on patterns, vulnerability can either be *persistent* or *situational* (Rashed and Weeks 2003b). For example, an individual can choose or be compelled to live in a hazardous location because of lack of access to resources and power, which is common in developing countries (Haque 1997). This can be termed as situational. Persistent vulnerability is a social construct that largely originates from socioeconomic and cultural deprivations. The vulnerability of individuals or communities is changeable with respect to time and space (Cutter et al. 2003; Parker 2000). They could attempt to minimize losses from an extreme event by exercising relevant mitigation measures based on their prior experience of hazards.

Despite the disparity in terminology, causations, and techniques, literature suggest that assessment of vulnerability to natural hazards has considerable implications for society. First, vulnerability assessments can extensively contribute to disaster risk reduction (Gao et al. 2007). Second, they promote the development of a disaster-resilient society (Birkmann 2006). They also assist in developing mitigation measures at local, national, and international levels (Alexander 2000; Cutter 1996, 2003a; White and Haas 1975) and are useful in identifying factors that result in vulnerable conditions (Yoon 2012; Cutter et al. 2009; Meyer et al. 2009; Rashed et al. 2007; Barroca et al. 2006; Chakraborty et al. 2005; Pelling 1997, 2002, 2003; Uitto 1998; Hewitt 1997; Watts and Bohle 1993; Anderson 1992).

Hazard is defined as a threat that can potentially cause damage to people, property, or other elements. It can be natural (earthquake), technological (chemical spill), or man-made (civil war) (Godschalk 1991). Hazards may be characterized by location,

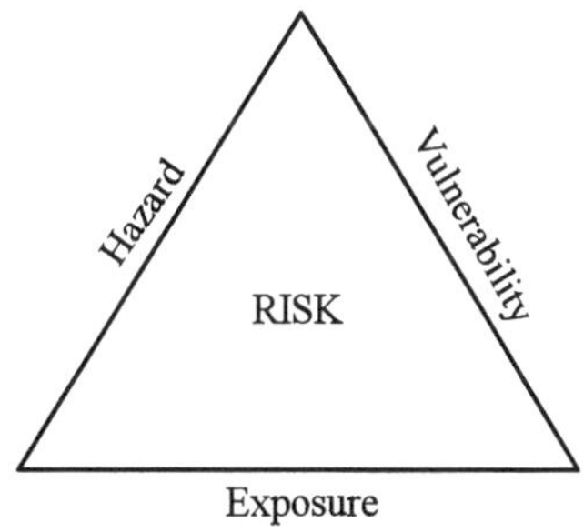

Fig. 2.1 Risk triangle (Adapted from Crichton 2002)

time, intensity, and frequency (Schneiderbauer and Ehrlich 2004; Wisner et al. 2004; Smith and Ward 1998). Typical characteristics of flood hazards include area of inundation, flood depth, frequency, rainfall–runoff lag times, and geomorphological settings (Few 2003; Alcántara-Ayala 2002; Weng 2001; Parker 2000).

Risk is viewed as the probability of occurrence or the degree of loss of a specified element expected from a specific hazard (Schneiderbauer and Ehrlich 2004). While risk measurement differs according to discipline, in hazard research, risk is equal to the product of two or three factors (Crichton 2002, 2007; Kohler et al. 2005; ADRC 2005; Kron 2005; Wisner et al. 2004; UN/ISDR 2004), although different views exist (Chakraborty et al. 2005). For example, Crichton (2002) illustrates risk with a triangle in which hazard, exposure, and vulnerability contribute independently (see Fig. 2.1). Conversely, Asian Disaster Reduction Center [ADRC] (2005) describes risk as the overlapping areas of three factors—hazard, exposure, and vulnerability—that act simultaneously to generate the risk of natural hazards, which can be expressed as:

$$\text{Risk} = \text{hazard} \times \text{vulnerability} \tag{2.1}$$

$$\text{Risk} = \text{hazard} \times \text{exposure} \times \text{vulnerability} \tag{2.2}$$

While hazards are a potential threat to populations and the environment, risk is the interplay between hazard and vulnerability. *Elements at risk*, a frequently used term in hazard research, enables the estimation of economic losses from an extreme event (Meyer et al. 2009). It is usually not included in the risk equation; rather, it is part of the vulnerability and exposure analysis (Fedeski and Gwilliam 2007). However, according to the United Nations Disaster Relief Coordinator Office (United Nations Development Programme [UNDP] 1992), risk is the function of elements at risk (e.g., population), hazards, and vulnerability. It differs from the concept of others, who define risk as a product of hazard and vulnerability (see Wisner et al. 2004). The risk to a particular community varies over time and time and depends on their socioeconomic, cultural, and other attributes (Wisner et al. 2004; Cannon 2000), signifying that the risk of natural hazards depends on both the hazard and the capability of the community to withstand shocks from disaster. To determine the level of vulnerability of a community or individuals to natural

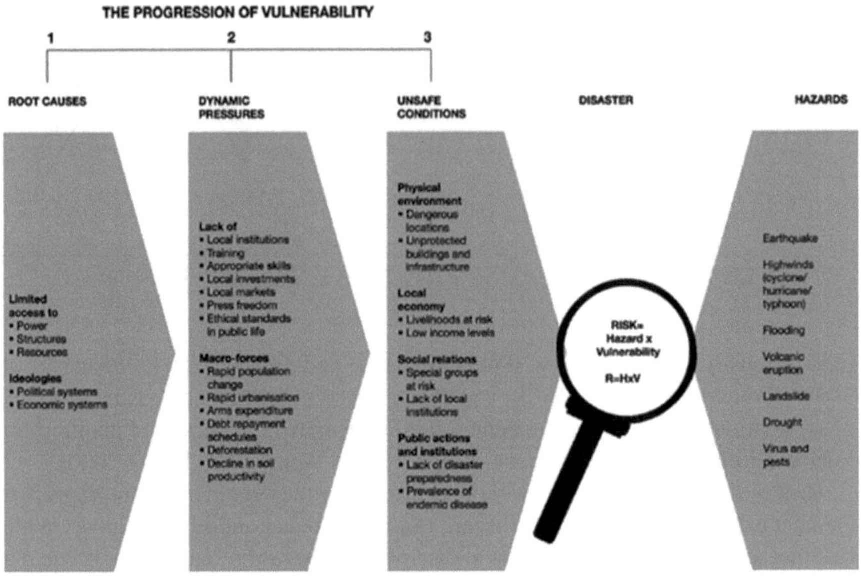

Fig. 2.2 Pressure and release model (*PAR*) (Adapted from Wisner et al. 2004)

hazards, it is important to inspect coping capacity elements while analyzing risk posed by natural hazards (UNDP 1992).

2.3 Models of Vulnerability and Disaster Risk Reduction

A number of conceptual models are available to typify hazards, vulnerability, and risk. The following sections present some of the frequently used theories to conceptualize a working framework for present study.

2.3.1 Pressure and Release (PAR) Model and Access Model

The PAR model (see Fig. 2.2) defines vulnerability as '*the characteristics of a person or group and their situation that influence their capacity to anticipate, cope with, resist and recover from the impact of a natural hazard* (Wisner et al. 2004, p. 11),' and risk is the product of hazard and vulnerability (Wisner et al. 2004; Blaikie et al. 1994). This theoretical framework serves to analyze exposure to a natural hazard and vulnerability—focusing on the process and causality of human vulnerability—and to explain how a naturally occurring phenomena turns into a disaster. Vulnerability is generated by three elements: root causes, dynamic pressures, and unsafe conditions. Root causes are the economic, demographic, and political processes that differentiate people in terms of their access to resources such as food, information, and economic systems. Dynamic pressures are economic, social, and political activities or processes that transform the effects of the root causes into the vulnerability of unsafe conditions such as a lack of appropriate skills, rapid population growth, and a decline in soil productivity. Unsafe conditions are the specific locations or situations where dangerous locations, low-income levels, and a lack of disaster preparedness can place a particular community or individuals at risk of natural hazards. Disaster occurs when two opposing forces intersect, that is, processes that generate vulnerability intersect with exposure to a hazard (Wisner et al. 2004). Although the model is able to tract the progression of vulnerability, it fails to take into account the coupled human–environment system associated with proximity to a hazard (Cutter et al. 2008). Moreover, many dynamic pressures and unsafe conditions are likely to be highly dependent on the local level rather than the global and national levels (Birkmann 2006).

The access model is an extended version of the PAR model, and it mainly focuses on how vulnerability is developed in cyclic order at the micro scale, such as households. The model further indicates that vulnerability is not independent of hazards and that it is changeable with respect to time and situation. In this model, the key to attenuate or amplify human vulnerability is *access to resources*, which is essential to recover from a disaster and mitigate the future risk of natural hazards. Although physical vulnerability or exposure to a hazard is evenly distributed in an area, individuals that have access to more resources have a higher capability of recovering from a disaster relatively early, whereas people with fewer opportunities and capabilities tend to exhibit elevated vulnerability.

2.3.2 *Hazards-of-Place (HOP) Model*

Borrowing an idea from Hewitt and Burton on the hazardousness of places, Cutter (1996) posits the *hazards-of-place* (HOP) model, which integrates biophysical and social characteristics relating to the causal process of hazards. Cutter asserts that biophysical and social elements collectively influence the vulnerability of a place. In addition, they can be employed to examine single and multi-hazards, and they are applicable to any geographic unit such as local, regional, or global.

The basic premise of this model is that vulnerability is related to interacting elements, such as location and the people living there, and the interaction and intersection of social and biophysical elements. It defines vulnerability as the combination of a social construct (social vulnerability) and a biophysical condition (potential exposure) in a spatial context. Therefore, social and biophysical conditions equally influence the vulnerability of a place, which should be investigated separately. Risk is defined as the likelihood of the occurrence of a particular hazard (e.g., a 100-year flood), and it is seen as an element of *place vulnerability*.

In this model, hazard potential is produced by a combination of risk and mitigation measures. Hazard potential is examined in a geographical context, such as elevation and proximity to a hazardous source, to recognize biophysical or

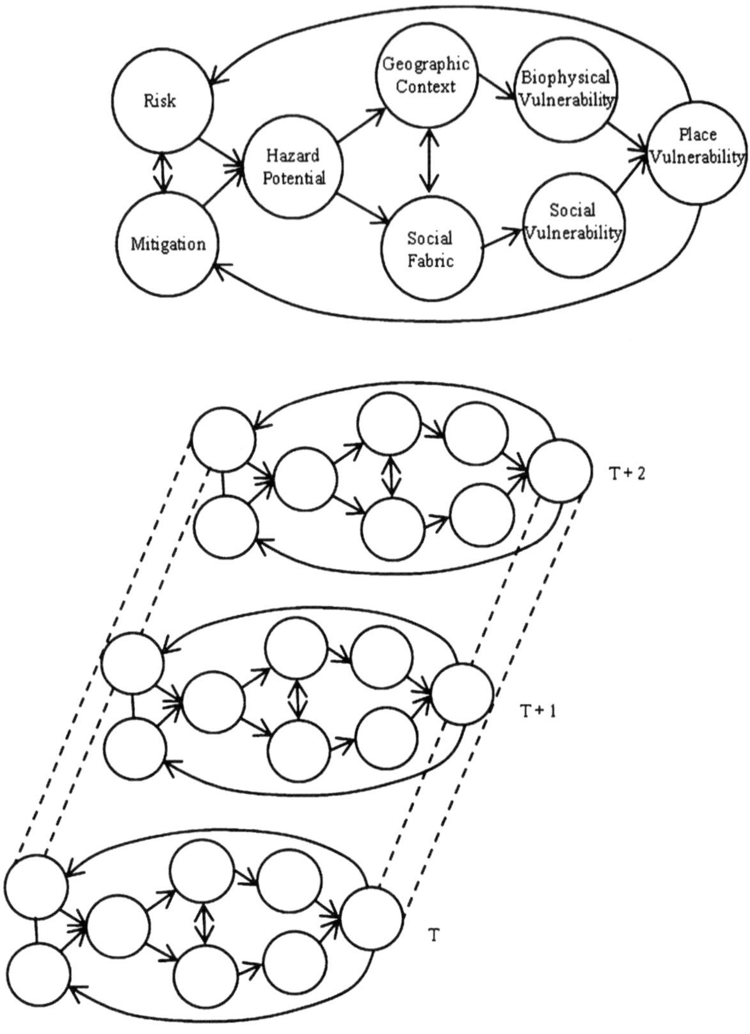

Fig. 2.3 Hazards-of-place model (Adapted from Cutter 1996)

environmental vulnerability, and, in a social context, to identify social vulnerability. The two vulnerabilities are then intersected to create the vulnerable places within a specific geographic domain. This process is repeated over time and provides the vulnerability as feedback to risk and mitigation at the start of the process. Place vulnerability can change over time, as efforts to mitigate hazards may alter the risk of a particular community by taking appropriate measures (see Fig. 2.3). An important pitfall of this model is that biophysical and social indicators are given equal weight. Further, it does not take into account the root causes of social vulnerability, larger contexts, and post-disaster effects and recovery (Cutter et al. 2008). An extended version of the HOP model has recently been proposed by Cutter

Fig. 2.4 Conceptual framework for disaster risk reduction (Adapted from Davidson 1997)

et al. (2008), which they called *the disaster resilience of place (DROP)* model. Although the DROP model was originally developed for natural hazard study, it is equally applicable to global change research.

In addition to these two popular models, there are a few other models that characterize the vulnerability of a community or individuals to environmental hazards. While it is beyond the scope of this book to review them all here, some of the models are discussed further below.

2.3.3 Regions of Risk Model

In his *regions of risk* model, Hewitt (1997) suggests that vulnerability arises from four broad principles. First, a hazard is a potential threat that humans want to avoid. Second, vulnerability, or adaptability, refers to the characteristics of a community or an individual. Depending on the attributes, potential damage can be high or low for a given hazard. Third, risk depends neither on hazard nor vulnerability; however, intervening conditions can influence the risk of hazards. For example, elevation can contribute to the occurrence of flood and thus act as an intervening factor, which Hewitt (1997) termed as *contingent vulnerability*. The fourth principle comprises people's coping and adjustment capacities to a dangerous environmental setting. Both capabilities depend on the social organization of a community or individuals' capacity to respond to an extreme condition.

2.3.4 Disaster Risk Reduction Model

Vulnerability is considered as part of the disaster risk reduction approach, and risk is the product of hazard, exposure, vulnerability, and coping capacity (Davidson 1997). In this model, Davidson proposed four broad frameworks that exhibit the disaster risk of a given community (see Fig. 2.4). First, hazard is the probability or severity of an event. Second, exposure characterizes structure, population, and economy. Third, vulnerability encompasses physical, social, economic, and environmental aspects. Fourth, capacity and mitigation measures include physical planning, social capacity, economic capacity, and management. Using these four measures, it is possible to determine community's vulnerability to hazards and take the necessary actions to lessen the risk of disaster. This model was mainly developed to estimate the urban earthquake disaster risk index; hence, this is applicable to single hazard investigations.

2.3.5 Integrated Assessment of Multi-hazards Model

Using the concepts of Blaikie et al. (1994) and Hewitt (1997), Greiving et al. (2006) proposed a new model called the *integrated assessment of multi-hazards*. This model ascribes a spatial component, as every hazard must take place over space. They considered the assessment of risk rather than vulnerability, and risk is seen as the product of hazard and vulnerability. The model is based on three important principles. First, a multi-hazard perspective is used rather than a single hazard. Second, it is only applicable for hazards that have spatial relevance, such as an earthquake. Spatially non-relevant hazards such as disease epidemics or traffic accidents cannot be used. Third, the model may be useful to determine community risk by integrating hazard and vulnerability; however, it is unable to recognize individuals' risk. An important pitfall of this model is that it requires copious data to operationalize the concept.

2.3.6 The UNISDR Framework for Disaster Risk Reduction

UNISDR developed a framework for disaster risk reduction in which vulnerability is the key to determine the risk of a community. ISDR (2004) classified vulnerability into four categories: physical, social, economic, and environmental. The pathways that are placed in the framework are comprehensive and able to distinguish different phases of disaster risk reduction techniques, including vulnerability assessment, hazard identification, assessment of risk, and early warning and response. However, while the approach failed to explain how vulnerability reduction could lead to the reduction of risk, an important contribution to this framework is linking vulnerability and disaster risk reduction elements in the context of *sustainable development* (Birkmann 2006).

Most of the models, particularly PAR and HOP, have been tested in different environments to investigate societal risk and vulnerability to natural hazards (Rashed et al. 2004; Kiunsi et al. 2006; Bollin and Hidajat 2006; Boruff et al. 2005; Wu et al. 2002; Mustafa 1998; Clark et al. 1998; Longhurst 1995). Although the results of these studies are laudable, data scarcity, particularly in developing countries, could be a serious issue to effectively use either of the models described above (Wang et al. 2011; Kiunsi et al. 2006). For example, the PAR model is

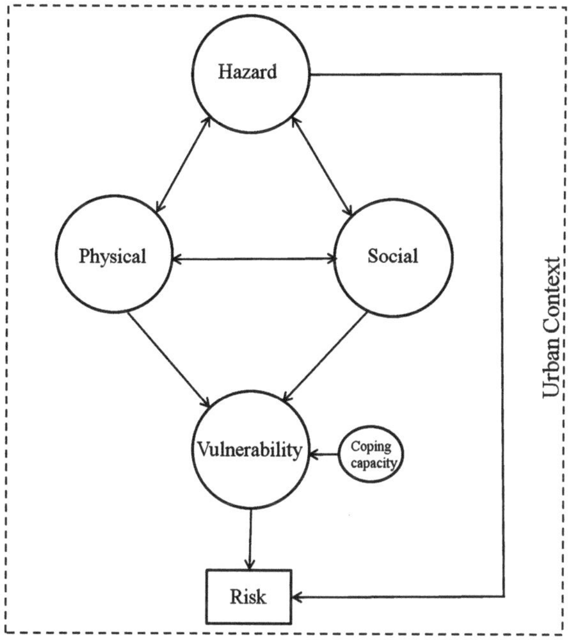

Fig. 2.5 Schematic representation of the conceptual framework

valuable for descriptive analysis rather than empirical testing, while the HOP approach allows the concept to be operationalized through geospatial techniques (Cutter et al. 2009). As a result, one must contextualize the indicators and methods of evaluation on the basis of both the spatial and socioeconomic milieu of the study area (Birkmann 2007; Mitchell 1999a).

2.4 Conceptual Framework

As the study is intended to examine the risk of natural hazards at community level, many of the existing models that are usually applied to the western world are of little relevance to this study. In addition, an urban environment is a complex and dynamic phenomenon; therefore, adopting a single approach may not be adequate to the deeper understanding of vulnerability to natural hazards in Dhaka, where lack of data is an inherent problem. For instance, a political ecological approach does not provide a methodological basis to integrate biophysical data within a geographic information system (GIS), although it is superior in identifying social differences. Consequently, the framework presented here is primarily based on the integrative notion of Cutter's (1996) HOP theory. However, the framework also borrowed elements from Mitchell et al. (1989), UNDP (1992), Hewitt (1997), Davidson (1997), Wisner et al. (2004), and Greiving et al. (2006) and embedded them in a spatial domain (see Fig. 2.5). The major elements of the framework are described below.

A hazard is viewed as a threat that has the potential to overwhelm people, property, and the environment. It is a pre-existing condition that can lead to devastation depending on the influence of external and internal forces; however, the magnitude of an event is strongly related to human and economic loss. External drivers may include global phenomena such as El Niño, which could lead to abnormalities in the regular functioning of natural systems. In contrast, internal drivers are primarily regional and/or local, and they mainly result from anthropogenic activities. For example, changes in land use lead to the changes in the hydrological cycle, which significantly affects flooding. Other internal drivers, such as rapid urban expansion, population growth, existing policies, economic conditions, and sociopolitical segregation, may force people to settle in dangerous locations. External and internal drivers can act independently or in combination to make a particular hazard condition worse. For instance, intense rainfall (external) combined with a dilapidated drainage system (internal) can upset an urban system relatively quickly.

Exposure to hazard is treated as a given and depends on internal drivers, where a number of factors, such as scarcity of land, inappropriate legislation, and political structure, compel people to settle in hazardous areas, as they have limited options. However, they may take countermeasures beforehand to lessen potential threats; for instance, people in flood-prone areas usually raise their houses from the surrounding floodplain. This is an example of a pre-emptive measure that could attenuate and amplify the immediate effects of a disaster. Thus, exposure to hazard is implicit and is not part of vulnerability because humans wish to avoid it. As there is inequality in resources and power structures, people have very few options to settling in hazardous locations. It should be noted that some drivers are measureable and others are not. Peak flow, flood extent, velocity, and depth of flooding can be used to depict the spatial variation of flood hazard in a given area.

Vulnerability is seen as the interplay between the social and physical aspects of a particular system (e.g., urban), and it is location-specific. The concept of vulnerability combines elements from physical and social spaces that make a particular system or its subsystem susceptible to experiencing loss and injury from natural hazards. A community's vulnerability can also be influenced by exogenous and endogenous factors. For example, globalization is an exogenous factor that may influence urban vulnerability to environmental hazards. Contrary to some conceptualizations, the model also views that overall vulnerability depends on the physical and social attributes, as well as on the coping capacity, of a given community. To effectively determine the total vulnerability of a community, the existing coping capacity should be taken into account. Physical vulnerability refers to the attributes of natural and built environments, such as elevation, geomorphic constructs, land-use practice, and proximity to hazard site. Social vulnerability, which embraces demographics, socioeconomic, infrastructures, and lifelines, defines the vulnerability of people, particularly marginal groups. Coping capacity is a form of adaptive capacity, which is perceived as the capability of a community to withstand or recover from disasters. Since the study area is flood-prone, it is reasonable to assume that communities are aware of the flood risk and are prepared to face disaster. Literacy, experience of past events, income, and availability of critical facilities such as hospitals may serve as indicators of coping capacity. Affiliations with social organizations may also be incorporated to measure the coping capability of a community.

The model stresses that risk is the product of hazard and vulnerability. To lessen the effects of natural hazards, steps should be taken to reduce vulnerability, which could lessen damage and ultimately the risk of natural hazards. For instance, regulating land-use practice could alleviate vulnerability, or building code practices could assist in reducing earthquake damage.

Although the model is exploratory in nature and applies to flood, the proposed framework can be used for other hazards, provided that the area of investigation presents similar types of attributes. In multi-hazard assessments, it is advisable to estimate vulnerability for each hazard (e.g., cyclone) separately and subsequently combine each vulnerability map. However, indicators for risk and vulnerability assessments may vary according to the scale of analysis. The proposed framework applies to collective risk (e.g., community) rather than individual risk. An important point to consider is that all elements can vary with time depending on the mitigation practices of an area or community. To implement this framework, one must consider several criteria, including the single or multi-hazard perspective and spatially relevant phenomena. The model is able to support all approaches, including inductive, deductive, and multi-criteria-based techniques. Further, geospatial techniques can be employed with advanced spatial analytical methods to assist in testing the framework empirically. In addition, it is indicative, suggesting that researchers could adapt hazard, vulnerability, and risk indicators depending on the area and hazard being considered.

2.5 Vulnerability and Risk Assessment with Geospatial Techniques

Geospatial techniques[1] such as geographic information system (GIS) and remote sensing (RS) show immense potential for the management of natural hazards, as many of the critical problems associated with extreme events are inherently spatial (Kienberger 2012; Morelli et al. 2012; Rashed et al. 2007; Cutter 2003b; Zerger and Smith 2003; Rashed and Weeks 2003a; Zerger 2002; Cova 1999; Coppock 1995; Bruce 1994; Emmi and Horton 1993; Carrara and Guzzetti 1995; Rejeski 1993). While GIS allows spatial decision making (Cova 1999) through the integration and analysis of spatial and aspatial data, RS is efficient in providing instantaneous and synoptic views essential for assessing the magnitude of extreme events spatially (Cutter 2003b; Ramsey et al. 2001). A large synoptic view with repeat coverage of satellites is known to be beneficial for mapping and monitoring of natural hazards (Gillespie et al. 2007; Sanyal and Lu 2004). Due to its spatial analytical capability, GIS is predominantly able to answer questions related to space, while RS data can aid

[1] Geospatial techniques refer to the suite of geographical information systems, remote sensing, global positioning systems (GPS), and spatial analysis.

in development planning (Imhoff et al. 1987). Satellite data are particularly useful for developing countries, as it is costly and time consuming for governments to update their databases with the traditional ground-based method due to lack of resources (Dong et al. 1997). Hence, the use of geospatial techniques in disaster management is evolving rapidly, as evidenced by scientific research over the past few years (Showalter and Lu 2010; Taubenböck et al. 2009; van Oosterom et al. 2005; Sanyal and Lu 2004; Showalter 2001).

Recently, the integrating capabilities of RS with GIS have opened up new opportunities for the quantitative analysis of natural hazards at all geographic and spatial scales. The integrated use of GIS and RS could lead to a deeper understanding of complex natural hazards in the spatial and temporal context, which is regarded as vital for disaster and/or emergency management, particularly the management of flood risk (Meyer et al. 2009; Mansor et al. 2004; Coppock 1995). In addition, GIS-based risk and vulnerability analysis can assist in the identification of human–hazard interactions (Rashed et al. 2007; Rashed and Weeks 2003a).

To understand how geospatial techniques are being utilized in emergency management, one needs to understand the different phases of the disaster cycle (Johnson 1992). Therefore, it is worth describing *the cycle of disaster* before embarking on how these techniques are playing a crucial role in assessing natural hazard risk. A disaster management cycle (Alexander 1993, 2000) or an emergency management system (Drabek and Hoetmer 1991; Godschalk 1991) primarily includes four cyclical segments: mitigation, preparedness, response, and recovery. However, these four phases can be compressed to three (mitigation, preparedness and response, and recovery) while using GIS in emergency management (Cova 1999). As GIS is capable of identifying objects spatially, it can be efficiently employed in each of the emergency management phases that are deemed suitable in order to enhance one's knowledge of risks caused by natural hazards (Gunes and Koval 2000). A brief account on the utility of geospatial techniques in emergency/disaster management is provided here, and further detail can be found in Cova (1999) and Cutter (2003b).

Mitigation Phase
Mitigation can be defined as the efforts to reduce the risk of life and property losses from natural and man-made hazards. In this step, geospatial techniques can be exploited in many ways. For example, emergency managers can develop long-term assessment, planning, and management of a particular event using geospatial data so that spatial variation in hazard, vulnerability, and risk can be determined and mapped (Cova 1999; Bocco et al. 1995; Johnson 1992; Watson 1992). To develop pertinent mitigation measures in relation to an event, hazard identification, risk mapping, and/or assessment are valuable instruments to which spatial analysis can contribute considerably.

Preparedness Phase
In contrast, disaster preparedness refers to activities such as operational capabilities that relate to emergency response (Gunes and Koval 2000). To prepare against a natural hazard, geospatial techniques can be used to delineate evacuation routes (Dunn and Newton 1992), assess vulnerability (Kienberger 2012; Yoon 2012; Finch

et al. 2010; Bhattarai and Conway 2010; Fekete et al. 2010; Taubenböck et al. 2008, 2009; Azar and Rain 2007; Gao et al. 2007; Rashed et al. 2007; Boruff et al. 2005; Montz and Tobin 2003; Rashed and Weeks 2003a; Cutter 2003b; Wu et al. 2002; Odeh 2002; Hill and Cutter 2001; Cutter et al. 2000; Clark et al. 1998), aid potential evacuation difficulties (Cova and Church 1997), and map evacuation needs (Chakraborty et al. 2005). As effective communication can significantly reduce risks from potential threatening events, mapping accurate hazard areas, locating critical facilities, mobilizing resources during and after an event, and physical constraint mapping of the incident site are key areas where geospatial techniques can be exploited (Dymon 1990). Hazard modeling based on historical events may also contribute effective risk communication to a community for a particular hazard (Cova 1999).

Response Phase
In contrast to disaster preparedness, the response phase mainly deals with minimizing loss and saving lives during and after an event. This stage also includes rescue and relief operations. As noted by Cutter (2003b), GIS-based incident command systems and consequence analysis tools can greatly assist emergency managers to respond immediately after a disaster. Another noteworthy application of geospatial techniques in this phase could be the map of available services (e.g., electricity, transport) to the public in the event of a disaster (Cutter 2003b).

Recovery Phase
The final stage of the emergency or disaster management cycle is known as the recovery phase, which includes damage assessment, reconstruction, and assistance to the disaster victims. Overall, the goal of this stage is to restore normal life in disaster-affected areas. An initial priority of the phase is to perform a cursory damage assessment to minimize the time needed to conduct the relief operation (Difani and Dolton 1992). In addition to rapid damage assessment and relief work, geospatial techniques can also extensively assist in rebuilding, assuaging and educating people, coordinating the spatial inventory of damage, and preventing reoccurrence (Gokon and Koshimura 2012; Wang and Li 2008; Sirikulchayanon et al. 2008; Cova 1999). Nevertheless, distributed spatial information in the form of web mapping techniques is another potential area that can significantly enhance people's capabilities for saving lives and property from natural disasters (Tate et al. 2011).

As described above, geospatial techniques can be applied to any phase of a disaster cycle as well as any natural or man-made hazards, including flood, earthquake, tornado, cyclone, landslide, and fire. Although it is beyond the scope of this work to demonstrate how geospatial techniques provide important information for mitigating risks from various natural hazards, a literature review on the application of geospatial techniques to hydrological assessments could be important to discern the uniqueness of these techniques for flood-related risk assessments. Additionally, it would allow the identification of the pathways needed to operationalize the conceptual framework described above. Rather than explaining the utility of these techniques in flood risk management using the four disaster phases listed above, for simplicity, the discussion is restricted to three distinct areas because some of the phases overlap, making it difficult to allocate studies to a particular phase (Cova 1999). The three broad categories in which geospatial approach plays a critical role in flood risk management are listed below. The first category presents different techniques of flood delineation using optical and microwave data, and the subsequent sections describe how geospatial techniques support flood damage and risk assessment:

1. Flood mapping and monitoring
2. Damage assessment
3. Risk assessment, including hazard and vulnerability estimation

2.5.1 Flood Mapping and Monitoring

RS data from earth-observing satellites provide the instantaneous and synoptic view necessary to accurately map the extent of floods. Therefore, it is widely used to delineate inundation areas, and it is considered highly potential to reduce flood losses and deaths (Corbley 1993). At present, optical, microwave, and laser RS are the major data sources for hydrologic application; however, the main hindrance of the optical sensor is that its imaging capability is restricted during a flood due to cloud coverage, particularly in tropical countries. To overcome this drawback, a Synthetic Aperture Radar (SAR) sensor has been offered an important avenue to the hydrologic community (McMillan et al. 2006; Tralli et al. 2005; Smith 1997). In addition to its all-weather imaging capability, SAR data are predominantly helpful for monsoon countries where stormy weather and cloud cover preclude imaging by optical sensors (Sanyal and Lu 2004; Smith 1997; Imhoff et al. 1987). Furthermore, SAR is believed to be effective in detecting lowland floods where subtle topographic variations inhibit the demarcation of precise flood mapping using a digital elevation model (Townsend 2001). In contrast, laser RS such as airborne light detection and ranging (LIDAR) also showed great potential to estimate flooding (Webster et al. 2004, 2006); however, the application of LIDAR is largely confined to developed countries due to its cost.

The majority of the previous works that used optical RS on floods utilized the multispectral scanner (MSS) on ERTS-1 (later renamed Landsat). It is found that two methods are primarily applied to delineate the flood extent using optical data. In the first method, threshold-based image segmentation was used, where a flood time image could be divided into two categories (flooded and non-flooded) using the provided threshold value (Barrett and Curtis 1992). In the second method, digital image analysis techniques were used, which are popularly known as supervised and unsupervised techniques. In this method, information from several bands could be used simultaneously (Engman and Gurney 1991).

Landsat MSS was used extensively to map the flood extent, mainly in the USA (Rango and Solomon 1974; Moore and North 1974; Williamson 1974; Ackleson et al. 1985; Imhoff et al. 1987; Pietroniro and Prowse 1996). The MSS band7 (0.8–1.1 μm) proved to be the most capable in discriminating water from dry soil or vegetated surface, as water strongly absorbs the near-infrared portion of the electromagnetic spectrum (Smith 1997). Although errors in MSS-derived flood areas were estimated at less than 5% (Rango and Solomon 1974), only 64% accuracy was reported by Imhoff et al. (1987), who failed to adequately resolve the raised dike system. Moore and North (1974) reported a larger error for flooded forests. In the case of forested wetlands, flooding could be efficiently mapped when the vegetation was dormant or in a leafless condition (Carter 1982). Under other conditions, the spectral reflectance of vegetation was used as an indicator of the presence of floodwater beneath the canopy (Ormsby and Mulligan 1985). Further, the presence of suspended materials in the water or floating vegetation increases the water's spectral value, which could create difficulties in distinguishing flooded and non-flooded areas (Engman and Gurney 1991).

To determine flood-prone areas, Landsat Thematic Mapper (TM) and MSS were used concurrently in a study by Nagarajan et al. 1993, who demonstrated that, by identifying river course changes between two dates, flood-prone lands could be detected accurately. Baumann (1996) used two TM4 (before and during flood) images in his 1993 Mississippi River flood analysis. Using the density-slicing technique, Baumann (1996) classified images into water, flooded, and non-flooded areas. Due to some confusion, he combined TM7 with TM4 to overcome the perplexity between water and industrial areas. In another study, Baumann (1999) used TM band4 to map the flood extent but experienced difficulties in separating water from certain urban features without the inclusion of an additional band.

Recent Landsat TM studies revealed that change-detection algorithms such as image differencing, normalized difference change detection (NCCD) (Gianinetto and Villa 2011), image ratio, principal component analysis (PCA), tasseled cap transformation, and change vector analysis can contribute significantly to delineating and detecting flooding in diverse environments. For example, Wang et al. (2002) used Landsat TM to identify water and non-water areas. In their study, TM4 was added to TM7, and normalized difference vegetation index or NDVI-based thresholding method was used. Threshold values were provided based on water reflectance and other categories that were determined by ground truth and image histogram.

The potential of multi-temporal Landsat TM using four change-detection algorithms was also examined to detect areas affected by flood and erosion in Nepal (Dhakal et al. 2002). The study revealed that the visible band was more accurate in identifying affected areas. Of the four methods used, spectral change vector analysis (SCVA) was found to be suitable in detecting areas affected by flood and erosion. Wang (2004) employed Landsat 7 to identify water versus non-water areas. Various combinations of TM bands, including band ratios, differences, and additions, were examined. He then classified TM data into three distinct categories: water body, flooded, and non-flooded areas. The overall accuracy of the flood maps was between 82.5 and 99%. In addition, automated approaches such as the normalized difference water index (NDWI) or tasseled cap transformation are also used to delineate flood-prone areas using Landsat data; however, success with these techniques largely depends on the area under study (Jain et al. 2005).

Satellite Pour l'Observation de la Terre (SPOT) data were also extensively used to determine flooded areas. Lorenz et al. (2001) demonstrated that the stereo capabilities of SPOT provide a unique opportunity for hydrologic modeling. The near-infrared channel of SPOT produces very low reflection in water, which is useful in separating water and non-water areas (de Brouder 1994). Blasco et al. (1992) used nine SPOT XS data to delineate the flood extent. Visual and digital analysis indicated that the extracted flooded areas were four times larger compared with the dry seasonal hydrographic network. Flood stage mapping and changes in channel geometry and patterns were also studied using the SPOT multispectral data (van Westen 1993). A spectral band ratio was employed to separate water and land areas. Extensive flooding may have a significant effect on the vegetation community, which can be analyzed with RS. A study by Michener and Houhoulis (1997) suggested that storm-induced flooding seriously affected forest ecosystems in Italy. Using five unsupervised change-detection algorithms, the responses of vegetation during flood were identified by creating an NDVI image. As noted by Lawal et al. (2011), NDVI-based change detection is prone to the misclassification of flooded areas because the index is influenced by other factors, including atmospheric state; therefore, the use of NDVI in flood mapping should be done very carefully (Sanyal and Lu 2004).

Due to its capability to image considerably larger areas at a low cost (Colwell and Hicks 1985), advanced very high-resolution radiometer (AVHRR) data are an important data source in studying flood (Smith 1997). Wiesnet et al. (1974) used NOAA-2 VHRR (later renamed AVHRR) to estimate the 1973 Mississippi River flood. Ali et al. (1989) demonstrated success in estimating the 1984 Bangladesh flood by comparing AVHRR visible, near-infrared, and thermal infrared channels. Barton and Bathols (1989) found that an AVHRR nighttime image delineated water bodies precisely in the case of the 1988 Darling River flood. Using a knowledge-based radial basis function neural network (RBFNN) model, Zhou et al. (2000) considered multi-temporal AVHRR data to map flood dynamics in China. They further incorporated Radarsat data to compare the flooded areas obtained from the AVHRR data. The tonal variation of images and the local knowledge of areas being studied also play a crucial role in determining flooded areas with RS data. A near-infrared band of AVHRR was used to separate water (flood) areas from non-water (land) categories, as the reflection of near-infrared by water is low and appears as a dark tone in the images (Rasid and Pramanik 1990).

Islam and Sado (2000a, b) employed multi-temporal flood time images of AVHRR to study flood hydrology. Using supervised and iterative self-organizing data analysis technique (ISODATA), their study revealed that the unsupervised classification method produced convincing results in obtaining flooded areas compared to supervised classification. All of the aforementioned methods require cloud-free AVHRR images; otherwise, inundation mapping may not be possible. To overcome the cloud contamination problem with an AVHRR image, Islam and Sado (2000b) developed an algorithm to remove the cloud. The application of the algorithm showed a significant improvement in mapping flooded areas (Islam and Sado 2000b). Sheng et al. (1998) also developed a method to remove the effects of semi-transparent clouds and cloud shadows that are inherent in AVHRR images. They considered spectral characteristics, satellite signal components, and the main groundcover types to estimate flooded areas. New-generation sensors such as the moderate resolution imaging spectroradiometer (MODIS) also shows enormous potential to study floods, primarily due to its high temporal resolution (Islam et al. 2010; Sakamoto et al. 2007).

Although the majority of the optical RS studies utilized Landsat, SPOT, and AVHRR images to study flooding, relatively few works have used high-resolution optical sensors such as IRS-1C/1D Pan, LISS-III and IKONOS, QUICKBIRD, WORLDVIEW, and GEOEYE data. Due to the cost, few studies have exploited these high-resolution images to study flood (Srivastava et al. 2000; van der Sande et al. 2003; Ebert et al. 2009). However, products from these images were used to assess the post-flood damage and recovery effort (see Ebert et al. 2009).

Optical sensors onboard cannot image the earth due to clouds and other atmospheric disturbances, which are considered significant impediments for flood and hydrological mapping (Rango and Anderson 1974). The innovation of microwave RS, particularly radar RS, resolves the problem associated with optical sensors, although microwave data may produce different types of errors (Palmann et al. 2008; Dewan et al. 2006; Taft et al. 2003; Töyora et al. 2002). However, radar RS is a viable alternative data source to traditional optical sensing (Foody 1988), particularly for monsoon countries (Imhoff et al. 1987). As early as 1967, data from the radar systems were important for detecting and mapping water bodies because of their sensitivity to moisture conditions (Waite et al. 1981). In addition to its all-weather capability, SAR images have become the most familiar approach to studying flooding and associated problems around the world (Palmann et al. 2008). Another key advantage of using SAR data is that land and water can easily be distinguished (Töyora et al. 2002).

Various space- and airborne radar data have been used and tested to map the flood extent in versatile environments. Initially, supervised image classification method was used to classify SAR images into discrete land-cover types (Hoque et al. 2011; Dewan et al. 2006; Malnes et al. 2002; Töyora et al. 2002; Adam et al. 1998; Leconte and Pultz 1991). In addition, the threshold technique is the most commonly employed approach to separate inundated areas from non-inundated areas (Dewan and Yamaguchi 2008; Brivio et al. 2002; Liu et al. 2002; Townsend 2001; Zhou et al. 2000; Zhou and Qing 1999). The threshold value is provided empirically or statistically by the analyst and can be determined from the radar backscatter value, known as decibel (dB). A different value for land-cover classification is provided and tested simultaneously, and the empirical values are usually determined based on the area under investigation and the spectral properties of the image. Selecting the optimal threshold value is the critical part of this method, as the provided threshold value can significantly influence the accuracy of the output. Consequently, the accuracy of maps is determined using different statistical indices. Fung and LeDrew (1988) suggested that the value that produces the highest Kappa coefficient could be used as the optimal cutoff. In other circumstances, an adaptive threshold value could be utilized, which assumes a Gaussian distribution of the data (Ober et al. 1998 in Brivio et al. 2002). Liu et al. (2002) used time-series SAR to determine flood boundaries with the threshold technique but utilized a method similar to maximum-value composite (MVC) to correct the logical errors of classified SAR images. Image differencing and multi-temporal enhancement techniques were used to detect flood boundaries and examine the correlation between radar backscatter and water level changes in hurricane-related flooding in Louisiana (Kiage et al. 2005). The study observed a strong relationship between water level and radar backscatter; however, the presence of vegetation and the variation in elevation substantially influenced the radar signature.

Visual interpretation or change-detection algorithms that require two SAR images (before and after the flood) also exhibit promising techniques in estimating floods. Visual interpretation consists of developing a false color composite (FCC) by employing multi-temporal SAR data. The flooded land can then be drawn on a computer screen using the FCC image. This simple technique can be used to determine the progress of flooding (Long and Trong 2001). Principal component analysis (PCA) is also employed to delineate flood boundaries if multi-temporal data are available (Henebry 1997 in Brivio et al. 2002). This method offers great opportunities for detecting flooded areas, and the threshold value can be obtained from the standard deviation of the PCA image. Image textural information was also exploited to detect flood boundaries from radar images (Schumann and Baldassarre 2010; Solbø et al. 2004; Solbø and Solheim 2004; Horritt et al. 2001). Recent advances in the coherence and amplitude technique of SAR images have added a new dimension to detecting inundation areas (Sanyal and Lu 2004). For example, in the coherence approach, areas are generally identified as flooded where the coherence or correlation of radar backscatters from before and after flood imagery is very low. In the amplitude approach, areas are regarded as flooded where the radar backscatter value is in considerable decline from the before to after flood images (Nico et al. 2000). An Advanced Synthetic Aperture Radar (ASAR) image from Envisat was used in flood mapping during the Elbe River flood in 2002, and it reported that like- and cross-polarization data were highly beneficial in estimating flood in the context of crisis mapping (Henry et al. 2006). New-generation radar data such as TerraSAR-X is also used to estimate urban flooding in the UK through semiautomated algorithm (Mason et al. 2010). In another study, Mason et al. (2007) suggested a method to improve accuracy of the flood extent maps by incorporating airborne laser altimetry data with SAR.

Data from various SAR instruments such as those on Seasat—the Shuttle Imaging Radar missions (SIR-A, SIR-B, SIR-C)—and Japanese Earth Resources Satellites-1 (JERS-1) were extensively used to map flooding in forested floodplains and forested wetlands (Townsend and Walsh 1998; NASA 1998; Hess et al. 1995; Hess and Melack 1994; Imhoff and Gesch 1990; Ormsby et al. 1985; Richards et al. 1987). Most of these studies demonstrated that longer wavelengths, such as an L band with horizontal–horizontal (HH) polarization, are useful for detecting floods under forests. As the wavelength is longer than the leaf size, it can easily penetrate the forest canopy (Wang et al. 1995). However, a considerable enhancement of the scattering coefficient from flooded forests was also reported (Richards et al. 1987; Ormsby et al. 1985). To resolve this problem, Wang et al. (1995) suggested incorporating four components (diffuse scattering from the forest floor, volume scattering from the canopy, scattering from the canopy to the forest floor and specular reflectance back to the sensor, and scattering from the trunk to the floor and specular reflectance to the sensor) while employing the radar backscatter model using L band microwave data. Indeed, the modeling effort of canopy backscattering from flooded and non-flooded forests for C, L, and P bands by Wang et al. (1995) showed that L band (HH polarization) is superior to C band vertical–vertical (VV) polarization. Their study also showed that the C band with HH polarization shows potential for mapping flooded areas. Other attempts show that the C band SAR with HH polarization can detect forest flooding in leaf-off conditions (Adam et al. 1998; Leconte and Pultz 1991). Conversely, C band VV data from ERS 1/2 (European Remote-Sensing Satellite) were not useful for detecting floods under forests (Kasischke and Bourgeau-Chavez 1997). However, a thorough investigation is needed to confirm these issues. Some argue that the like-polarized scene is superior to the cross-polarized scene (Crevier and Pultz 1996).

The utilization of microwave data on inundation mapping has also encountered a few problems in conjunction with some success (Palmann et al. 2008). One problem is associated with radar wavelengths and terrain characteristics. Generally, water bodies in SAR images appear in a dark tone that results in a smooth black tone due to their specular reflection. In contrast, rough surfaces produce a bright return to the radar antenna, resulting in a light grey to white tone (Dong et al. 1997). However, turbid water with emergent vegetation can significantly increase radar backscatter, which creates problems in accurately deciding flood boundaries (Smith 1997). Windy weather is another problem of imaging by radar sensor during flood, as wind can produce ripples in the water surfaces, which are deemed obstacles in efficiently analyzing SAR data (Chilar et al. 1992). Another pitfall of using SAR data relates to separating flooded settlements from non-flooded settlements in the presence of trees. As trees can enhance radar backscatter, inundation within settlements is very difficult to identify (Oberstadler et al. 1997). However, many studies confirmed that SAR data provided by the Radarsat is more useful than the ERS sensor due to its capability to image the earth's surface at various incidence angles (Sanyal and Lu 2004).

The information acquired by active sensors is different compared to passive sensors (Ulaby et al. 1981). Optical data comprises the reflectivity of the different spectral bands of objects, while radar data consists of the electrical properties of the object. For example, optical images provide data on vegetation types, whereas microwave images provide moisture content as well as structure. As a result, these two types of images complement each other. To take advantage of this, a combined approach for flood analysis received greater attention, as this approach showed profound outcomes in land-use mapping (Solberg et al. 1994), flood boundary delineation (Töyora et al. 2002; Zhou and Qing 1999; Imhoff et al. 1987), and crop discrimination (Brisco et al. 1989). Skidmore et al. (1986) used Landsat MSS and SIR-B data to classify different forest types with varying water levels in Australia and found that the merging of two sensors provided higher classification accuracy and superior results. Imhoff et al. (1987) employed SIR-B and Landsat MSS data for flood boundary delineation. They delineated the flood extent from individual sensors and then merged both datasets to facilitate flood damage assessment. Zhou and Qing (1999) employed Landsat TM and Radarsat SAR to accurately estimate flooded areas in a mountainous environment in China. Their study developed a technique to overcome the spuriously demarcated flooded area in SAR imagery by using TM data. Töyora et al. (2002) used SPOT and Radarsat data to map wetland flooding in Canada and noticed that they were less accurate as single sensors and produced far better results when used in combination. Using geographic object-based image analysis techniques, a recent study showed immense potential for flood estimating by combining optical and microwave data (Mallinis et al. 2011).

In addition to utilizing RS data alone for flood mapping, the integration of elevation data with RS is also used in several studies (Brivio et al. 2002; Wang et al. 2002) to demonstrate the contribution of geospatial techniques in hydrological assessments. These studies revealed that an integrated technique could provide more accurate information on inundation if quality data are available. For example, using a least accumulative cost–distance matrix and ERS-SAR data, Brivio et al. (2002) confirmed that flooded surface could be estimated more precisely even if a time delay in RS acquisition hinders real-time flood mapping during an event. The combined use of the digital elevation model (DEM) and Landsat data revealed that the DEM assisted in identifying flooding beneath forest canopies (Wang et al. 2002) and provided better results compared to Landsat-derived flood information. The DEM was also employed to obtain flood depth. Although grid size does have an impact on the calculation of flood depth (Werner 2001), a moderate-fine resolution DEM could be of significant value in hydrological modeling such as flood depth and damage estimation (Marwade et al. 2008; Gianinetto et al. 2006; Mark et al. 2004; de Brouder 1994).

2.5.2 *Flood Damage Assessment*

Data from earth-observation satellites with GIS have been extensively used to estimate flood damage. It involves estimating the effects of an event on different sectors such as agriculture (Yamagata and Akiyama 1988) and urban. Two Landsat data, along with digital image analysis in the form of a change-detection matrix and a Delta NDVI, were used to quantify agricultural damage in relation to flooding in Indiana (Pantaleoni et al. 2007). The location of villages and land-cover data were overlaid on SAR-derived flood maps to estimate flood damage in Thailand (Mongkolsawat and Thanajaturon 2006). The study corroborated that geospatial techniques can be an excellent alternative to conventional hydrological flood damage estimations, particularly for countries lacking detailed spatial data. Using slope data derived from a DEM with four sequential Landsat imageries, a method was developed to estimate flood damage in Italy (Gianinetto et al. 2006). An overall accuracy of 85.6% was achieved to demonstrate the efficacy of this method.

Land-cover data from a high-resolution image may allow accurate flood damage assessments (van der Sande et al. 2003). This study suggested that a deeper understanding of flood hazard is possible if a land-cover map from a high-resolution image is superimposed on the flood map obtained from DEM. A synthetic depth–damage curve was also prepared using the flood extent map, land-use, and flood depth data from the LISFLOOD simulation model. Crop and structural damage is another area where geospatial techniques are of great value (Sadars and Tabuchi 2000; Srivastava et al. 2000). For example, Profeti and Macintosh (1997) used pedological and land-use data with a flood extent map derived from a SAR image to determine damage statistics. The study concluded that a GIS could easily be used to quantify flood damage for crop and urban structures. Based on Radarsat and a 30-m DEM, a *flood-level-determination* algorithm was developed to assess damage to paddy fields (Waisurasingha et al. 2008). The study noted that a good-resolution DEM could significantly contribute to the accurate estimation of rice damage, even with a single-date flood time image.

Srivastava et al. (2000) determined infrastructural flood damage by combining IRS wide-field sensor-derived (WiFS) flood maps with infrastructural information within a GIS. A transport network database was overlaid on flood maps obtained from an AVHRR image to estimate road damage, and the output was reclassified to understand the risk of national and regional highways in Bangladesh (Islam and Sado 2000c). The dynamics of flood damage were estimated using multi-temporal flood maps with land-use/land-cover data to determine the degree of flood loss according to different land-use/land-cover categories (Dewan et al. 2005; Liu et al. 2002). These studies showed that geospatial techniques are important in determining flood damage in different environments. However, data that comprise satellite images taken before, during, and after the flood can allow the spatial variability of damage to be estimated, if they are available.

2.5.3 Flood Hazards, Vulnerability, and Risk Assessment

Geospatial techniques, particularly the spatial analytic capability of GIS, provide a number of advantages over traditional methods in natural hazard risk assessment (Wadge et al. 1993). Therefore, they are increasingly being used to assess the risk of naturally occurring phenomena around the world. The geospatial community commonly employs diverse data to evaluate flood risk and/or vulnerability. There are several techniques that assess flood risk using geospatial and hydrological data, which can broadly be divided into quantitative and qualitative/semiquantitative methods (Wang et al. 2011). The first type uses hydraulic information and terrain models to derive flood hazards with exceedance probability. The output is then integrated with land-use or property information to derive flood loss functions essential to determining flood risk (Su et al. 2005). The second type typically employs gauge records, satellite data, or both along with the elevation model to derive flood extent and depth. Information such as elements at risk (e.g., population density, buildings) can subsequently be integrated within a GIS to develop spatial coexistence models for flood risk assessment (Sanyal and Lu 2004). The quantitative method is able to provide superior results, but it requires fine-resolution DEM with advanced mathematical skills, while the qualitative and/or semiquantitative method is relatively easy to implement within a GIS, and most importantly, it allows compromising of data scarcity (Wang et al. 2011). Literature suggest that the majority of flood risk research, particularly in developing countries, is based on the latter type. However, the former is popular within the hydrological community (Zerger and Wealands 2004). A reasonable explanation is that quality data is lacking in developing countries; hence, RS products are a viable alternative for these countries. For example, flood depth is an important hydraulic parameter in flood risk modeling. A DEM is usually used to derive this variable; however, in the absence of fine-resolution elevation data, which is common in developing countries, satellite data can be of substantial help to calculate flood depth based on the turbidity of floodwaters (Sanyal and Lu 2005; Islam and Sado 2000, 2000a, b). Further, advances in image processing techniques are currently allowing scientists to derive the required information quickly, thereby supporting the parameterization of flood risk calculations in a given area.

Rejeski (1993) noted three empirical approaches to evaluating flood hazards using a GIS. One of the simplest approaches is the binary model, in which a hazard is expressed as either present or absent for a particular cell. The second approach, called a weighted (ordinal) model, includes ranking locations according to the severity of the hazard. The final approach is the quantitative (interval/ratio) model, in which users assign values to locations in space that could be used to compute the unit hazard factor.

Hydraulic data and elevation information, together with weighted spatial coexistence models, were used to derive flood hazards for various return periods; subsequently, land-use information was overlaid on each of the flood maps to generate a series of binary maps depicting the area and land-use category affected (Boyle et al. 1998). Flood depth information obtained from the elevation data was then used to approximate the depth–damage curve for Ontario, Canada. Similar works can be found in other countries to identify vulnerable and risk areas based on loss estimation (Suriya and Mudgal 2012; Kubal et al. 2009; Fedeski and Gwilliam 2007; Büchele et al. 2006; Su et al. 2005; van der Sande et al. 2003; Dutta et al. 2003, 2005; Nyarko 2002; Green 2000; Smith 1981). For example, flood extent and depth data were derived from a two-dimensional hydraulic model with a DEM to calculate flood risk in Hong Kong. Based on the flood depth, flood risk maps were generated (Brimicombe and Bartlett 1996). Using a topological overlay, Simonovic (1993) distinguished flooded areas from non-flooded areas in a GIS and then employed a quantitative model to obtain the flood damage function. Stochastic techniques and Monte Carlo simulations were also used to develop flood risk scenarios for the Rhine River in Germany. A number of models were developed to estimate river runoff scenarios in order to understand the effects of upstream levee breaches on downstream flood risk. Finally, the model estimated risk by multiplying flood probability with a depth–damage function (Apel et al. 2006). Hydrological simulation based on hydrodynamic models was used to assess flood hazards. The derived flood hazard map was subsequently integrated with population/land-use data to estimate flood risks within a GIS (Gain and Hoque 2012; Masood and Takeuchi 2012; Morita 2011; Büchele et al. 2006; Tingsanchali and Karim 2005; Zhang et al. 2003). Flood data from RS with a statistical technique such as logistic regression was also used to predict floods and associated risk (Pradhan 2009; Chubey and Hathout 2004).

The risk of death by natural hazard is considerably higher in communities that are socially disadvantaged (Thomas and Mitchell 2001; Cutter et al. 2003). Therefore, identifying communities that are vulnerable to natural hazards can significantly contribute to disaster risk reduction. Although vulnerability is a complex notation, geospatial techniques can help derive biophysical variables that invariably support vulnerability analysis of a given hazard. For example, flood possesses a significant threat to people living in low-lying areas or in the proximity of rivers. Elevation can thus be an important indicator to evaluate physical vulnerability to flood in a particular region. Land use/cover is another critical factor that can be derived quickly from satellite data to be incorporated into a hazard model.

In addition to biophysical variables, demographic and socioeconomic data can be obtained from the census and integrated into a GIS to assess social vulnerability (Cutter et al. 2009). Although the temporal resolution of census data is low and collect for different purposes (Ebert et al. 2009), it remains one of the most important data sources for developing countries for vulnerability mapping. However, several recent studies (Ebert et al. 2009; Rashed et al. 2007; Rashed and Weeks 2003a) showed that RS could be an excellent source to obtain the required data for the generation of vulnerability map. For example, an NDVI was used as an indicator of wealth to assess urban vulnerability to earthquakes (Rashed and Weeks 2003a, b). Similarly, Ebert et al. (2009) derived as many as 47 proxy variables from high-resolution satellite and other spatial databases to calculate social vulnerability in Honduras. The latter study further used a stepwise regression technique to identify the most relevant proxies. Whatever the data sources are, vulnerability to hazards is known to be a function of both physical (e.g., slope) and socioeconomic (e.g., income) elements (Cutter et al. 2009; Cova 1999).

A number of studies have used biophysical and socioeconomic variables to evaluate flood hazards and vulnerability in diverse environments (Kienberger 2012; Wang et al. 2011; Kazmierczak and Cavan 2011; Sebastian et al. 2011; Scheuer et al. 2011; Chen et al. 2011; Schumann and Baldassarre 2010; Bizimana and Schilling 2010; Pandey et al. 2010; Fekete et al. 2010; Finch et al. 2010; Intarawichian and Dasananda 2010; Pavri 2010; Maantay et al. 2010; Bizimana and Schilling 2010; Meyer et al. 2009; Kubal et al. 2009; Zheng et al. 2009; Zahran et al. 2008; Sinha et al. 2008; Demirkesen et al. 2007; Azar and Rain 2007; Sanyal and Lu 2006; Yalcin and Akyurek 2004; Montz and Tobin 2003; Jain and Sinha 2003; Wu et al. 2002; Cutter et al. 2000; Clark et al. 1998). As noted by Cutter et al. (2009) and Rashed and Weeks (2003b), the combined use of social and physical variables could allow a rigorous assessment of vulnerability to natural hazards, as they are intrinsically linked.

A range of techniques is available within the GIS to evaluate hazards and vulnerability in relation to floods. One of the common approaches to quantify flood vulnerability and risk is to construct an index by combining various indicators (Yoon 2012). For example, a modified z-score was used to develop a composite vulnerability index in relation to cyclone-related hazards (Islam 2008) and flood-related hazards (Zahran et al. 2008). Although the technique is simple and easy to implement in a GIS, the key variables that influence vulnerability in a region are not well understood. However, the results from this method are useful for comparing the spatial variability of vulnerability across different spatial units.

Pandey et al. (2010) used flood and waterlogging data with demographic and socioeconomic variables to estimate flood and waterlogging hazards and the vulnerability index at the district level in India. Using a rating scale (1–4), they generated waterlogging and flood hazard maps. Similarly, a composite vulnerability map was prepared by integrating various thematic maps derived from socioeconomic indicators. Finally, hazard and vulnerability scores were multiplied to obtain the risk of flood/waterlogging in the area. A weighted summation method was used to generate flood hazard maps for different administrative units for West Bengal, India, which comprised flood occurrence and socioeconomic data (Sanyal and Lu 2006).

Similar methods were used for flood hazard assessments at regional and local levels, and a flood risk map was created for different administrative units considering flood depth and flood frequency as hydrological parameters (Dewan et al. 2006, 2007; Sanyal and Lu 2005; Tanavud et al. 2004; Islam and Sado 2000, 2000a, b). The household-level flood risk was also mapped out in a recent study (Tran et al. 2009), underscoring the utility of geospatial techniques in the reduction of flood risk at the community level. Many studies have used weighted index and related approaches to derive flood hazard, vulnerability, and risk maps within a GIS because these approaches are easy to implement using the inbuilt capabilities of the GIS.

Spatial multi-criteria decision assessment/analysis (MCA) or multi-criteria evaluation (MCE) has received renewed interest because of the following: (1) it allows improved decision making (Malczewski 2004, 2006); (2) it supports developing and evaluating alternative plans (Malczewski 1996); and (3) it is predominantly appropriate for spatial decision making, as the data that the decision makers rely on are mostly related to space (Malczewski 1999). Even though the recent advancement of spatial information systems allows the integration of diverse data with MCA/MCE methods, few researchers have employed this technique for water resource management, particularly flood risk assessment (Meyer et al. 2009). However, several studies considered the MCA or related approach in the evaluation of flood mitigation measures (Haque et al. 2012; Akter and Simonovic 2005; Willett and Sharda 1991). These studies are based on field surveys or secondary data to evaluate flood risk alleviation; however, they did not account for the spatial component (Kubal et al. 2009). The analytic hierarchy process (AHP), which is one of the variants of MCA (Greene et al. 2011), is used to determine flood risk (Dang et al. 2011; Gao et al. 2007). Using a range of physical and social variables, Sinha et al. (2008) derived a flood risk index in North Bihar, India.

Sebastian et al. (2011), Meyer et al. (2009), and Kubal et al. (2009) conducted an innovative approach regarding flood vulnerability by combining ecological, economical, and social data. The authors used GIS-based multi-criteria assessment techniques to evaluate flood risk for the Mulde River, Germany, and a software tool called FloodCalc was developed to support risk calculation with single and/or multi-criteria (Meyer et al. 2009). Yalcin and Akyurek (2004) compared three approaches of MCE to derive a composite flood vulnerability index. The first approach uses a Boolean overlay that takes the intersection or union of all factors with false or true values determined by the criteria threshold. The second approach inversely ranks the factor criteria in terms of the significance of flood causes to decide weights for the factors. The last approach utilizes AHP to determine weights for the indicators. In the second and third approaches, an integration of the weighted factor layers was conducted by considering the weighted linear combination (WLC). Different approaches of MCA within a GIS were also tested and used for natural hazard assessment, including flood (Intarawichian and Dasananda 2010; Yahaya 2008; Levy et al. 2007; Rahman and Saha 2007; Rashed and Weeks 2003a). These studies revealed that flood risk assessments based on MCA/MCE provide informed decision making at any spatial scale, and they are also able to remove *errors* inherent in qualitative weighting for the determination of flood risk areas (Dang et al. 2011).

In addition to the approaches discussed above, other techniques have been used to measure flood vulnerability. For instance, an entropy method was used to identify exposure, vulnerability, and resilience at the county level of the Huaihe River basin in China. This study derived a flood risk map by using social and biophysical vulnerability with an inverse-variance weighting approach (Zheng et al. 2009). Kienberger (2012) developed a technique by integrating the *geon* approach to determine social and economic vulnerability to floods in Mozambique, and Cutter et al. (2003) adopted an additive model of an equal factor weighting for the social vulnerability index at the county level in the USA. The use of an inductive approach such as factor analysis appears to be a good technique when dealing with many variables to construct social, economic, and physical vulnerability to natural hazards (Yoon 2012). A number of studies employed factor analysis to determine community vulnerability to natural hazards (Kazmierczak and Cavan 2011; Burton and Cutter 2008; Azar and Rain 2007; Boruff and Cutter 2007; Boruff et al. 2005). For example, the risk of surface water flooding in Greater Manchester was estimated using socioeconomic and biophysical variables, and the Principal Component Analysis (PCA) was used to identify underlying factors contributing to social vulnerability. Finally, spatial associations between hazards, vulnerability, and exposure were determined, and flood risk of urban communities was mapped (Kazmierczak and Cavan 2011). In addition, the PCA is found to be very useful to map the spatial and temporal changes of social vulnerability to natural hazards at national level (Cutter and Finch 2008).

Geospatial techniques can potentially help in other areas of disaster management. For instance, a GIS can be efficiently used in the formulation of nonstructural mitigation measures such as disaster preparedness planning (Johnson 1992), optimal location of flood shelters (Sanyal and Lu 2009), and evacuation planning (Chakraborty et al. 2005). Relatively little has been done on these issues; however, the advancement of a spatially interoperable database may provide more avenues to deal with informed decision making during emergency and crisis situations.

The above discussion suggests that geospatial techniques are invaluable in determining vulnerability and natural hazard risks, particularly flooding in any spatial context. However, as literature show, geospatial techniques can also lead to uncertainty and errors in hazard modeling. Although it is difficult to document them explicitly, some of the issues related to uncertainty, challenges, and limitations in relation to natural hazards and emergency management can be found elsewhere (Ebert et al. 2009; Zerger and Smith 2003; Rashed and Weeks 2003a; Zerger 2002; Coppock 1995).

2.6 Summary

This chapter commenced with the terminologies commonly used in hazard literature and then summarized the existing models of vulnerability, disaster, and risk of natural hazards. A conceptual framework was proposed, and a thorough review of the utility of geospatial techniques in disaster management was subsequently presented. Although the conceptual models discussed above comprise merits and demerits, it is worth noting that vulnerability assessments have become an integral part of hazard research. Vulnerability analysis can support the formulation of a resilient society. Therefore, an integrative approach that comprises hazards and vulnerability should be considered, so the risk of natural hazards can be estimated and communicated.

A growing body of literature suggests that geospatial techniques can significantly contribute to the deeper understanding of flood hazard, risk, and vulnerability. As described, these techniques are capable of adding value in each stage of the emergency management/disaster cycle through spatial analyses. The most important outcome of the geospatial approach is a series of maps encompassing the spatial distribution of hazard, risk, and vulnerability, which is known to provide better representation and visualization of human–environment interaction (Bhattarai and Conway 2010; Clark et al. 1998; Varnes et al. 1984 in Wadge et al. 1993). As hazards take place over space, the incorporation of spatial components into risk assessments could lead to informed disaster risk reduction and hence allow the development of a disaster-resilient community (UN 2005).

References

Ackleson SG, Klemas V, McKim HL, Merry CJ (1985) A comparison of SPOT simulator data with Landsat MSS imagery for delineating water masses in Delaware Bay, Broadkill River and adjacent wetlands. Photogramm Eng Remote Sens 51(8):1123–1129

Adam S, Wiebe J, Collins M, Pietroniro A (1998) Radarsat flood mapping in the Peace-Athabasca delta, Canada. Can J Remote Sens 24(1):69–79

Adger W (2006) Vulnerability. Glob Environ Chang 16(3):268–281

Akter T, Simonovic SP (2005) Aggregation of fuzzy views of a large number of stakeholders for multi-objective flood management decision making. J Environ Manage 77(2):133–143

Alcántara-Ayala I (2002) Geomorphology, natural hazards, vulnerability and prevention of natural disasters in developing countries. Geomorphology 47(2–4):107–124

Alexander D (1993) Natural disasters. UCL Press Limited, London

Alexander D (2000) Confronting catastrophe: new perspectives on natural disasters. Oxford University Press, New York

Ali A, Quadir DA, Huh OK (1989) Study of river flooding hydrology in Bangladesh with AVHRR data. Int J Remote Sens 10(12):1873–1891

Anderson MB (1992) Metropolitan areas and disaster vulnerability: a consideration for developing countries. In: Kreimer A, Munasinghe M (eds) Environmental management and urban vulnerability. The World Bank, Washington, DC, pp 77–92

Apel H, Thieken A, Merz B, Blöschl G (2006) A probabilistic modelling system for assessing flood risks. Nat Hazard 38(1–2):79–100

Asian Disaster Reduction Centre (ADRC) (2005) Total disaster risk management – good practice. ADRC, Kobe, Japan. Available at http://www.adrc.or.jp/

Azar D, Rain D (2007) Identifying population vulnerable to hydrological hazards in San Juan, Puerto Rico. GeoJournal 69(1–2):23–43

Bankoff G (2001) Rendering the world unsafe: 'vulnerability' as Western discourse. Disasters 25(1):19–35

Bankoff G, Frerks G, Hilhorst D (2004) Mapping vulnerability: disasters, development & people. Earthscan, London

Barrett EC, Curtis LF (1992) Introduction to environmental remote sensing, 3rd edn. Chapman and Hall, London

Barroca B, Bernardara P, Mouchel JM, Hubert G (2006) Indicators for identification of urban flood vulnerability. Nat Hazard Earth Syst Sci 6(4):553–561

Barton IJ, Bathols JM (1989) Monitoring floods with AVHRR. Remote Sens Environ 30(1):89–94

Baumann PR (1996) Flood analysis: 1993 Mississippi flood. Volume 4 of remote sensing core curriculum. URL http://umbc7.umbc.edu/~tbenja1/baumann/baumann.html

Baumann PR (1999) http://www.research.umbc.edu/tbenja1/baumann/mode2.html. Accessed 10 Feb, 1999

Bhattarai K, Conway D (2010) Urban vulnerabilities in the Kathmandu valley, Nepal: visualizations of human/hazard interactions. J Geogr Inf Syst 2(2):63–84

Birkmann J (ed) (2006) Measuring vulnerability to natural hazards: towards disaster resilient society. United Nations University Press, Tokyo

Birkmann J (2007) Risk and vulnerability indicators at different scales: applicability, usefulness and policy implications. Environ Hazard 7(1):20–31

Bizimana JP, Schilling M (2010) Geo-information technology for infrastructural flood risk analysis in unplanned settlement: a case study of informal settlement flood risk in the Nyabugogo flood plain, Kigali city, Rwanda. In: Showalter PS, Lu Y (eds) Geospatial techniques in urban hazard and disaster analysis. Springer, Dordrecht, pp 99–124

Blaikie P, Cannon T, Davis I, Wisner B (1994) At risk: natural hazards, people's vulnerability and disasters. Routledge, London

Blasco F, Bellan MF, Chaudhury MU (1992) Estimating the extent of floods in Bangladesh using SPOT data. Remote Sens Environ 39:167–178

Bocco G, Sanchez R, Riemann H (1995) GIS affects flood planning efforts. GIS World 7(3):32–35

Bollin C, Hidajat R (2006) Community-based risk index: pilot implementation in Indonesia. In: Birkmann J (ed) Measuring vulnerability to natural hazards: towards disaster resilient society. United Nations University Press, Tokyo, pp 271–289

Boruff BJ, Cutter SL (2007) The environmental vulnerability of Caribbean island nations. Geogr Rev 97(1):932–942

Boruff BJ, Emrich C, Cutter SL (2005) Erosion hazard vulnerability of US coastal counties. J Clim Res 21(5):932–942

Boyle SJ, Tsanis IK, Kanaroglou PS (1998) Developing geographic information system for land use impact assessment in flooding condition. J Water Resour Plan Manage 124(2): 89–98

Brimicombe AJ, Bartlett JM (1996) Linking geographic information systems with hydraulic simulation modelling for flood risk assessment: the Hong Kong approach. In: Goodchild MF, Steyaert L, Parks BO, Josnston CO, Maidment DR, Crane MP, Glendinning S (eds) GIS and environmental modelling: progress and research issues. Wiley, New York, pp 165–168

Brisco B, Brown RJ, Manore MJ (1989) Early season crop discrimination with combined SAR and TM data. Can J Remote Sens 15(1):44–54

Brivio PA, Colmbo R, Maggi M, Tomasoni R (2002) Integration of remote sensing data and GIS for accurate mapping of flooded areas. Int J Remote Sens 33(3):429–441

Bruce JP (1994) The fifth session of the IDNDR Scientific and Technical Committee. Nat Hazard Obs 18(4):6–7

Büchele B, Kreibich H, Kron A, Thieken A, Ihringer J, Oberle P, Merz B, Nestmann F (2006) Flood-risk mapping: contribution towards an enhanced assessment of extreme events and associated risks. Nat Hazard Earth Syst Sci 6(4):485–503

Burton C, Cutter SL (2008) Levee failures and social vulnerability in the Sacramento-San Joaquin Delta area. Nat Hazard Rev 9(3):136–149

Burton I, Kates R, White GF (1993) The environment as hazard, 2nd edn. Guilford Press, New York

Cannon T (2000) Vulnerability analysis and disasters. In: Parker DJ (ed) Floods, vol 1. Routledge, London, pp 45–55

Carrara A, Guzzetti D (eds) (1995) Geographical information systems in assessing natural hazards. Kluwer Academic Publishers, Dordrecht

Carter V (1982) Applications of remote sensing to wetlands. In: Johannsen CJ, Sanders JL (eds) Remote sensing for resource management. Soil Conservation Society of America, Ankeny, pp 284–300

Chakraborty J, Tobin GA, Montz BE (2005) Population evacuation: assessing spatial variability in geophysical and social vulnerability to natural hazards. Nat Hazard Rev 6(1):23–33

Chen Y-R, Yeh C-H, Yu B (2011) Integrated application of the analytical hierarchy process and the geographic information system for flood risk assessment and flood plain management in Taiwan. Nat Hazard 59(3):1261–1276

Chilar J, Pultz TJ, Gray AL (1992) Change detection with synthetic aperture radar. Int J Remote Sens 13(3):401–414

Chubey MS, Hathout S (2004) Integration of RADARSAT and GIS modelling for estimating future Red River flood risk. GeoJournal 59(3):237–246

Clark GE, Moser SC, Ratick SJ, Dow K, Meyer WB, Emani S, Jin W, Kasperson JX, Kasperson RE, Schwartz HE (1998) Assessing the vulnerability of coastal communities to extreme storms: the case of Revere, MA, USA. Mitig Adapt Strat Glob Chang 3(1):59–82

Colwell JE, Hicks DR (1985) NOAA satellite data: a useful tool for macro inventory. Environ Manage 9(6):463–470

Coppock JT (1995) GIS and natural hazards: an overview from a GIS perspective. In: Carrara A, Ad Guzzetti F (eds) Geographical information systems in assessing natural hazards. Kluwer Academic Publishers, Dordrecht, pp 21–34

Corbley KP (1993) Remote sensing and GIS provide rapid response for flood relief. Earth Obs Mag 2:28–30

Cova TJ (1999) GIS in emergency management. In: Longley PA, Goodchild MF, Maguire DJ, Rhind DV (eds) Geographical information systems, management and application. Wiley, Chichester, pp 845–858

Cova TJ, Church L (1997) Modeling community vulnerability evacuation using GIS. Int J Geogr Inf Sci 11(8):763–784

Crevier Y, Pultz TJ (1996) Analysis of C-band SIR-C radar backscatter over a flooded environment, Red River, Manitoba. In: Applications of remote sensing in hydrology, proceedings of the third international workshop, NHRI symposium no. 17, Greenbelt, MA, pp 47–60

Crichton D (2002) UK and global insurance response to flood hazard. Water Int 27(1):119–131

Crichton D (2007) What can cities do to increase resilience? Philos Trans R Soc 365(1860):2731–2739

Cutter SL (1996) Vulnerability to environmental hazards. Prog Hum Geogr 20(4):529–539

Cutter SL (2003a) The vulnerability of science and the science of vulnerability. Ann Assoc Am Geogr 93(1):242–261

Cutter SL (2003b) GI science, disasters, and emergency management. Trans GIS 7(4):439–445

Cutter SL, Finch C (2008) Temporal and spatial changes in social vulnerability to natural hazards. Proc Natl Acad Sci 105(7):2301–2306

Cutter SL, Mitchell JT, Scott MS (2000) Revealing the vulnerability of people and places: a case study of Georgetown county, South Carolina. Ann Assoc Am Geogr 90(4):713–737

Cutter SL, Boruff BJ, Shirley WL (2003) Social vulnerability to environmental hazards. Soc Sci Q 84(2):242–261

Cutter SL, Barnes L, Berry M, Burton C, Evans E, Tate E, Webb J (2008) A place-based model for understanding community resilience to natural disasters. Glob Environ Chang 18(4):598–606

Cutter SL, Emrich CT, Webb JJ, Morath D (2009) Social vulnerability to climate variability hazards: a review of literature. Final report to Oxfam America. Available at http://adapt.oxfamamerica.org/resources/Literature_Review.pdf. Accessed 2 Apr, 2011

Dang NM, Babel MS, Luong HT (2011) Evaluation of flood risk parameters in the Day river flood diversion area, Red River delta, Vietnam. Nat Hazard 56(1):169–194

Davidson R (1997) An urban earthquake disaster risk index. Report no. 121. Department of Civil Engineering, Stanford University, Stanford

de Brouder JAM (1994) Flood study in the Meghna-Dhonagoda Polder, Bangladesh. http://www.gisdevelopment.net/aars/acrs/1994/ts3/ts3007pf.htm. Accessed 10 May, 2003

Demirkesen AC, Evrendilek F, Berberoglu S, Kilic S (2007) Coastal flood risk analysis using Landsat-7 ETM+imagery and SRTM DEM: a case study of Izmir, Turkey. Environ Monit Assess 131(1–3):293–300

Dewan AM, Yamaguchi Y (2008) Effects of land cover changes on flooding: example from Greater Dhaka of Bangladesh. Int J Geoinf 4(1):11–20

Dewan AM, Yeboah KK, Nishigaki M (2005) Flood mapping and damage evaluation in Greater Dhaka, Bangladesh with remote sensing. Asian Profile 33(5):495–512

Dewan AM, Yeboah KK, Nishigaki M (2006) Using synthetic aperture radar (SAR) data for mapping river water flooding in an urban landscape: a case study of Greater Dhaka, Bangladesh. J Hydrol Water Resour 19(1):44–54

Dewan AM, Islam MM, Kumamoto T, Nishigaki M (2007) Evaluating flood hazard for land-use planning in Greater Dhaka of Bangladesh using remote sensing and GIS techniques. Water Resour Manage 21(9):2101–2116

Dhakal AS, Amada T, Aniya M, Sharma RR (2002) Detection of areas associated with flood and erosion caused by a heavy rainfall using multitemporal Landsat TM data. Photogramm Eng Remote Sens 68(3):233–239

Difani C, Dolton L (1992) Battling the Oakland fire with GIS. GIS World 5(1):43–45

Dong Y, Forster B, Ticehurst C (1997) Radar backscatter analysis for urban environments. Int J Remote Sens 18(6):1351–1364

Dow K (1992) Exploring differences in our common future(s): the meaning of vulnerability to global environmental change. Geoforum 23(3):417–436

Dow K, Downing TE (1995) Vulnerability research: where things stand. Hum Dimens Q 1:3–5

Drabek TE, Hoetmer GJ (eds) (1991) Emergency management: principles and practice for local government. International City Management Association, Washington, DC

Dunn CE, Newton D (1992) Optimal routes in GIS and emergency planning applications. Area 24(3):259–267

Dutta D, Herath S, Musiake K (2003) A mathematical model for flood loss estimation. J Hydrol 277(1–2):24–49

Dutta D, Khatun F, Herath S (2005) Analysis of flood vulnerability of urban buildings and population in Hanoi, Vietnam. Seisan Kenkyo 57(4):150–154

Dymon UJ (1990) The role of emergency mapping in disaster response. Quick response report no. 42. Natural Hazards Research and Applications Information Center, University of Colorado, Boulder

Eakin H, Luers A (2006) Assessing the vulnerability of social-environmental systems. Ann Rev Environ Resour 31:365–394

Ebert A, Kerle N, Stein A (2009) Urban social vulnerability assessment with physical proxies and spatial metrics derived from air- and spaceborne imagery and GIS data. Nat Hazard 48(2):275–294

Emmi PC, Horton CA (1993) A GIS-based assessment of earthquake property damage and causality risk: Salt Lake County, Utah. Earth Spectra 9(1):11–33

Engman ET, Gurney RJ (1991) Water resources management and monitoring. In: Engman ET, Gurney RJ (eds) Remote sensing in hydrology. Chapman Hall, London, pp 193–210

Fedeski M, Gwilliam J (2007) Urban sustainability in the presence of flood and geological hazards: the development of a GIS-based vulnerability and risk assessment methodology. Landscape Urban Plan 83(1):50–61

Fekete A, Damm M, Birkmann J (2010) Scales as a challenge for vulnerability assessment. Nat Hazard 55(3):729–747

Few R (2003) Flooding, vulnerability and coping strategies: local responses to a global threat. Prog Dev 3(1):43–58

Finch C, Emrich CT, Cutter SL (2010) Disaster disparities and differential recovery in New Orleans. Popul Environ 31(4):179–202
Foody GM (1988) Crop classification from airborne synthetic aperture radar data. Int J Remote Sens 9(4):655–668
Fung T, LeDrew E (1988) The determination of optimal threshold levels for change detection using various accuracy indices. Photogramm Eng Remote Sens 54:1449–1454
Gaillard JC, Liamzon CC, Villanueva JD (2007) Natural disasters? A retrospect into the causes of the late-2004 typhoon disaster in Eastern Luzon, Philippines. Environ Hazard 7(4):257–270
Gain AK, Hoque MM (2012) Flood risk assessment and its application in the eastern part of Dhaka city, Bangladesh. J Flood Risk Manage. doi: 10.1111/jfr3.12003
Gao J, Nickum JE, Pan Y (2007) An assessment of flood hazard vulnerability in the Dongting lake region of China. Lakes Reserv Res Manage 12(1):27–34
Gianinetto M, Villa P (2011) Mapping hurricane Katrina's widespread destruction in New Orleans using multisensor data and the normalized difference change detection (NCCD) technique. Int J Remote Sens 32(7):1961–1982
Gianinetto M, Villa P, Lechi G (2006) Postflood damage evaluation using Landsat TM and ETM+data integrated with DEM. IEEE Trans Geosci Remote Sens 44(1):236–243
Gillespie TW, Chu J, Elizabeth F, Thomas D (2007) Assessment and prediction of natural hazards from satellite imagery. Prog Phys Geogr 31(5):459–470
Godschalk DR (1991) Disaster mitigation and hazard management. In: Drabek TE, Hoetmer GJ (eds) Emergency management: principles and practice for local government. International City Management Association, Washington, DC, pp 131–160
Gokon H, Koshimura S (2012) Mapping of building damage of the 2011 Tohoku earthquake tsunami in Miyagi prefecture. Coast Eng J 54(1):1–12
Green C (2000) Handbook of water economics: principles and practice. Wiley, Chichester
Green C (2004) The evaluation of vulnerability to flooding. Disast Prev Manage 13(4):323–329
Greene R, Devillers R, Luther JE, Eddy BG (2011) GIS-based multiple-criteria decision analysis, Geogr Compass 5(6):412–432
Greiving S, Fleischhauer M, Luckenkotter J (2006) A methodology for an integrated risk assessment of spatially relevant hazards. J Environ Plan Manage 49(1):1–19
Gunes AE, Koval JP (2000) Using GIS in emergency management operations. J Urban Plan Dev 126(3):136–149
Haque CE (1997) Hazards in fickle environment: Bangladesh. Kluwer Academic Publishers, Dordrecht
Haque AN, Grafakos S, Huijsman M (2012) Participatory integrated assessment of flood protection measures for climate adaptation in Dhaka. Environ Urban 24(1):197–213
Henry J-B, Chastanet P, Fellah K, Desnos Y-L (2006) Envisat multi-polarized ASAR data for flood mapping. Int J Remote Sens 27(10):1921–1929
Hess LL, Melack JM (1994) Mapping wetland hydrology and vegetation with synthetic aperture radar. Int J Ecol Environ Sci 20:197–205
Hess LL, Melack JM, Filoso S, Wang Y (1995) Delineation of inundated area and vegetation along the Amazon floodplain with SIR-C synthetic aperture radar. IEEE Trans Geosci Remote Sens 33(4):896–903
Hewitt K (1997) Regions of risk: a geographical introduction to disasters. Longman, Essex
Hill AA, Cutter SL (2001) Methods for determining disaster proneness. In: Cutter SL (ed) American hazardscapes: the regionalization of hazards and disasters. The Joseph Henry Press, Washington, DC, pp 37–60
Hogan DJ, Marandola E Jr (2005) Towards in interdisciplinary conceptualization of vulnerability. Popul Space Place 11(6):455–471
Hoque R, Nakayama D, Matsuyama H, Matsumoto J (2011) Flood monitoring, mapping and assessing capabilities using RADARSAT remote sensing. GIS Ground Data Bangladesh 57(2):525–548
Horritt MS, Mason DC, Luckman AJ (2001) Flood boundary delineation from synthetic aperture radar imagery using a statistical active contour model. Int J Remote Sens 22(13): 2489–2507

Imhoff ML, Gesch DB (1990) The derivation of a sub-canopy digital terrain model of a flooded forest using synthetic aperture radar. Photogramm Eng Remote Sens 56(8):1155–1162
Imhoff ML, Vermillion C, Story MH, Choudhury AM, Gafoor A, Polcyn F (1987) Monsoon flood boundary delineation and damage assessment using spaceborne imaging radar and Landsat data. Photogramm Eng Remote Sens 53(4):405–413
Intarawichian N, Dasananda S (2010) Analytical hierarchy process for landslide susceptibility mapping in lower Mae Chaem watershed, Northern Thailand. Suranaree J Sci Technol 17(3):277–292
Islam T (2008) Cyclone wind analysis and disaster planning: an integrated approach for the Bangladesh coast. VDM Verlag, Saarbrücken
Islam MM, Sado K (2000a) Development of flood hazard maps of Bangladesh using NOAA AVHRR with GIS. Hydrol Sci J 45(3):337–355
Islam MM, Sado K (2000b) Flood hazard assessment in Bangladesh using NOAA AVHRR data with geographical information system. Hydrol Process 14(3):605–620
Islam MM, Sado K (2000c) Satellite remote sensing data analysis for flood damage zoning with GIS for flood management. Ann J Hydraul Eng (JSCE) 44:301–306
Islam MM, Sado K (2002) Development of priority map flood countermeasures by remote sensing data with geographic information system. J Hydrol Eng 7(5):346–355
Islam AS, Bala SK, Haque MA (2010) Flood inundation map of Bangladesh using MODIS time-series data. J Flood Risk Assess 3(3):210–222
Jain V, Sinha R (2003) A geomorphological manifestations of the flood hazards: a remote sensing based approach. Geocarto Int 18(4):51–60
Jain SK, Singh RD, Jain MK, Lohani AK (2005) Delineation of flood-prone areas using remote sensing techniques. Water Resour Manage 19(4):333–347
Jha AK, Bloch R, Lamond J (2012) Cities and flooding: a guide to integrated urban flood risk management for the 21st century. The World Bank, Washington, DC
Johnson GO (1992) GIS applications in emergency management. URISA J 4(1):66–72
Kasischke ES, Bourgeau-Chavez LL (1997) Monitoring south Florida wetlands using ERS-1 SAR imagery. Photogramm Eng Remote Sens 63(3):281–291
Kazmierczak A, Cavan G (2011) Surface water flooding risk in urban communities: analysis of vulnerability, hazard and exposure. Landscape Urban Plan 103(2):185–197
Kiage LM, Walker ND, Balasubramanian S, Babin A, Barras J (2005) Applications of Radarsat-1 synthetic aperture radar imagery to assess hurricane-related flooding of coastal Louisiana. Int J Remote Sens 26(24):5359–5380
Kienberger S (2012) Spatial modelling of social and economic vulnerability to floods at the district level of Buzi, Mozambique. Nat Hazard. doi:10.1007/s11069-012-0174-9
Kiunsi RB, Mechack V, Uhinga G, Mayunga J, Mulenge F, Kabali C, Sigalla N, Bilia M (2006) Disaster vulnerability assessment: the Tanzania experience. In: Birkmann J (ed) Measuring vulnerability to natural hazards: towards disaster resilient society. United Nations University Press, Tokyo, pp 227–245
Kohler A, Jülich S, Bloemertz L (2005) Guidelines risk analysis – a basis for disaster risk management, GTZ
Kron F (2005) Flood risk = hazard*values*vulnerability. Water Int 30(1):58–68
Kubal C, Haase D, Meyer V, Scheuer S (2009) Integrated urban flood risk assessment – adapting a multi-criteria approach to a city. Nat Hazard Earth Syst Sci 9(6):1881–1895
Lankao PR, Qin H (2011) Conceptualizing urban vulnerability to global change and environmental change. Curr Option Environ Sustain 3(3):142–149
Lawal DU, Matori A-N, Hashim AM, Chandio IA, Sabri S, Balogun A-L, Abba HA (2011) Geographic information system and remote sensing applications in flood hazards management: a review. Res J Appl Sci Eng Technol 3(9):933–947
Leconte R, Pultz TJ (1991) Evaluation of Radarsat for flood mapping using simulated satellite SAR imagery. Can J Remote Sens 17(3):241–249
Levy JK, Hartmann J, Li KW, An Y, Asgary A (2007) Multi-criteria decision support systems for flood hazard mitigation and emergency response in urban watersheds. J Am Water Resour Assoc 43(2):346–358

Lewis J (1999) Development in disaster-prone places: studies of vulnerability. Intermediate Technology Publications, London
Liu Z, Huang F, Liu L, Wan E (2002) Dynamic monitoring and damage evaluation of flood in north-west Jilin with remote sensing. Int J Remote Sens 23(18):3669–3679
Liverman DM (1990a) Vulnerability to global environmental change. In: Kasperson RE, Dow K, Golding D, Kasperson JX (eds) Understanding global environmental change: the contributions of risk analysis and management. Clark University, Worcester, pp 27–44
Liverman DM (1990b) Drought in Mexico: climate, agriculture, technology and land tenure in Sonora and Puebla. Ann Assoc Am Geogr 80(1):49–72
Long NT, Trong BD (2001) Flood monitoring of Mekong River delta, Vietnam using ERS SAR data. A paper presented at the 22nd Asian conference of remote sensing, Singapore, 5–9 November
Longhurst R (1995) The assessment of community vulnerability in hazard prone areas. Disasters 19(3):269–270
Lorenz N, Van Dijk M, Kwadijk J (2001) Use of remotely sensed images by SPOT in hydrologic modeling. In: Van Dijk M, Bos MG (eds) GIS and remote sensing techniques in land and water management. Kluwer Academic Publishers, Dordrecht, pp 39–53
Maantay J, Maroko A, Culp G (2010) Using geographic information science to estimate vulnerable urban populations for flood hazard and risk assessment in New York City. In: Showalter PS, Lu Y (eds) Geospatial techniques in urban hazard and disaster analysis. Springer, Dordrecht, pp 71–97
Malczewski J (1996) A GIS-based approach to multiple criteria group decision making. Int J Geogr Inf Syst 10(8):955–971
Malczewski J (1999) GIS and multiple-criteria decision analysis. Wiley, New York
Malczewski J (2004) GIS-based land-use suitability analysis: a critical overview. Prog Plan 62(1):3–65
Malczewski J (2006) GIS-based multi-criteria decision analysis: a survey of the literature. Int J Geogr Inf Sci 20(7):703–726
Mallinis G, Gitas IZ, Giannakopoulos V, Maris F, Tsakiri-Strati M (2011) An object-based approach for flood area delineation in a transboundary area using ENVISAT ASAR and Landsat TM data. Int J Digit Earth. doi:10.1080/17538947.2011.641601
Malnes E, Guneriussen T, Høgda KA (2002) Mapping flood-area by Radarsat in Vanjø, Norway. In: Proceedings of 29th international symposium on remote sensing of the environment, Buenos Aires, Argentina, 4–8 April
Mansor S, Shariah MA, Billa L, Setiawan I, Jabar F (2004) Spatial technology for natural risk management. Disast Prev Manage 13(5):364–373
Mark O, Weesakul S, Apirumanekul C, Aroonnet SB, Djordjevic S (2004) Potential and limitations of 1D modelling of urban flooding. J Hydrol 299(3–4):284–299
Marwade V, Cook A, Coonrod J (2008) GIS techniques for creating river terrain models for hydrodynamic modelling and flood inundation mapping. Environ Model Software 23(10–11):1300–1311
Mason DC, Horritt MS, Dall' Amico JT, Scott TR, Bates PD (2007) Improving river flood extent delineation from Synthetic Aperture Radar using airborne laser altimetry. IEEE Trans Geosci Remote Sens 45(12):3932–3943
Mason DC, Speck R, Devereus B, Schumann GJ-P, Neal JC, Bates PD (2010) Flood detection in urban areas using TerraSAR-X. IEEE Trans Geosci Remote Sens 48(2):882–894
Masood M, Takeuchi K (2012) Assessment of flood hazard, vulnerability and risk of mid-eastern Dhaka using DEM and 1D hydrodynamic model. Nat Hazard. doi:10.1007/s11069-011-0060-x
McMillan A, Morley JG, Adams BJ, Chesworth S (2006) Identifying optimal SAR imagery specifications for urban flood monitoring: a hurricane Katrina case study. Paper presented at 4th international workshop on remote sensing for post-disaster response, Magdalene College, University of Cambridge, Cambridge, UK, 25–26 September
Meyer V, Scheuer S, Haase D (2009) A multicriteria approach for flood risk mapping exemplified at the Mulde river, Germany. Nat Hazard 48(1):17–39

Michener WK, Houhoulis PF (1997) Detection of vegetation changes associated with extensive flooding in a forested ecosystem. Photogramm Eng Remote Sens 63(12):1363–1374
Mitchell JK (1999a) Natural disasters in the context of mega-cities. In: Mitchell JK (ed) Crucibles of Hazard: mega-cities and disasters in transition. United Nations University, Tokyo, pp 15–55
Mitchell JK (1999b) Megacities and natural disasters: a comparative analysis. GeoJournal 49(2):137–142
Mitchell JK, Devine N, Jagger K (1989) A contextual model of natural hazards. Geogr Rev 79(4):391–409
Mongkolsawat C, Thanajaturon T (2006) Predicting the recurrent flood extent and damage assessment in Chi watershed, northeast Thailand using multi-temporal Radarsat and Landsat data. In: Proceedings of the first international conference on science and technology for sustainable development of the Greater Mekong Sub-region, Khon Khen, Thailand, 15–16 August. Available at http://202.28.94.77/research/i_proceed/2549/7_Predicting.pdf
Montz BE, Tobin GA (2003) Hazardousness of the Tampa region: evaluating physical risk and socio-economic vulnerability. In: Proceedings of the applied geography conferences, 31, pp 380–388
Moore GK, North GW (1974) Flood inundation in the southeastern United States from aircraft and satellite imagery. Water Resour Bull 10(5):1082–1096
Morelli S, Segoni S, Manzo G, Ermini L, Catani F (2012) Urban planning, flood risk and public policy: the case of the Arno River, Firenze, Italy. Appl Geogr 34(1):205–218
Morita M (2011) Quantification of increased flood risk due to global climate change for urban river management planning. Water Sci Technol 63(12):2967–2974
Müller A (2012) Areas at risk – concept and methods for urban flood risk assessment: a case study of Santiago de Chile. Franz Steiner Verlag, Stuttgart
Mustafa D (1998) Structural causes of vulnerability to flood hazard in Pakistan. Econ Geogr 74(3):289–305
Mustafa D (2005) The production of an urban hazardscape in Pakistan: modernity, vulnerability, and the range of choice. Ann Assoc Am Geogr 95(3):566–586
Nagarajan R, Marathe GT, Collins WG (1993) Identification of flood prone regions of Rapti river using temporal remotely-sensed data. Int J Remote Sens 14(7):1297–1303
NASA (1998) SAR references. http://southport.jpl.nasa.gov/science/SAR_REFS.html. Accessed 30 July, 2003
Nico G, Pappalepore M, Pasquarieloo G, Refice A, Samarelli S (2000) Comparison of SAR amplitude and coherence flood detection methods-a GIS application. Int J Remote Sens 21(8):1619–1631
Nyarko BF (2002) Application of a rational model in GIS for flood risk assessment in Accra, Ghana. J Spat Hydrol 2(1):1–14
O'Brien K, Leichenko R, Kelkar V, Venema H, Aandahl G, Tompkins H, Javed A, Bhadwal S, Barg S, Nygaard L, West J (2004) Mapping vulnerability to multiple stressors: climate change and globalization in India. Glob Environ Chang 14(4):303–313
O'Brien K, Eriksen S, Nygaard LP, Schojolden A (2007) Why different interpretations of vulnerability matter in climate change discourses. Clim Policy 7(1):73–88
Oberstadler R, Honsch H, Huth D (1997) Assessment of the mapping capabilities of ERS-1 SAR data for flood mapping: a case study in Germany. Hydrol Process 11:1415–1425
Odeh DJ (2002) Natural hazards vulnerability assessment for statewide mitigation planning in Rhode island. Nat Hazard Rev 3(4):177–187
Oliver-Smith A (2004) Theorising vulnerability in a globalised world: a political ecological perspective. In: Bankoff G, Frerks G, Hilhort D (eds) Mapping vulnerability: disasters, development & people. Earthscan, London, pp 10–24
Ormsby JP, Mulligan PJ (1985) The effects of merging TM and A/C Radar on wetland classification. In: Proceedings of the NASA/JPL Aircraft SAR workshop, USA, pp 37–38
Ormsby JP, Blanchard BJ, Blanchard AJ (1985) Detection of lowland flooding using active microwave systems. Photogramm Eng Remote Sens 51(3):317–328

Palmann C, Mavromatis S, Hernandez M, Sequeira M, Brisco B (2008) Earth observation using radar data: an overview of applications and challenges. Int J Digit Earth 1(2):171–195

Pandey AC, Singh SK, Nathawat MS (2010) Waterlogging and flood hazards vulnerability and risk assessment in Indo Gangetic plain. Nat Hazard 55(2):273–289

Pantaleoni E, Engel BA, Johannsen CJ (2007) Identifying agricultural flood damage using Landsat imagery. Precis Agric 8(1–2):27–36

Parker DJ (2000) Floods, vol I and II. Routledge, London

Pavri F (2010) Urban expansion and sea-level rise related flood vulnerability for Mumbai (Bombay), India using remote sensed data. In: Showalter PS, Lu Y (eds) Geospatial techniques in urban hazard and disaster analysis. Springer, Dordrecht, pp 31–49

Pelling M (1997) What determines vulnerability to floods: a case study in Georgetown, Guyana. Environ Urban 9(1):203–226

Pelling M (2002) Assessing urban vulnerability and social adaptation to risk. Int Dev Plan Rev 24(1):59–76

Pelling M (2003) The vulnerability of cities. Earthscan, London

Pietroniro A, Prowse TD (1996) Environmental monitoring of the Peace-Athabasca Delta using multiple satellite data sources. In: Applications of remote sensing in Hydrology, proceedings of the third international workshop, NHRI symposium no. 17, Greenbelt, MD, pp 237–251

Polsky C, Neff R, Yarnal B (2007) Building comparable global change vulnerability assessments: the vulnerability of scoping diagrams. Glob Environ Chang 17(3–4):472–485

Pradhan B (2009) Flood susceptible mapping and risk area delineation using logistic regression, GIS and remote sensing. J Spat Hydrol 9(2):1–18

Profeti G, Macintosh H (1997) Flood management through Landsat TM and ERS SAR data: a case study. Hydrol Process 11:1397–1408

Rahman MR, Saha SK (2007) Flood hazard zonation – a GIS aided multi-criteria evaluation (MCE) approach with remotely sensed data. Int J Geoinf 3(3):25–35

Ramsey EW III, Hodgson ME, Sapkota SK, Nelson GA (2001) Forest impact estimated with NOAA AVHRR and Landsat TM data related to a predicted hurricane wind-field distribution. Remote Sens Environ 77(3):279–292

Rango A, Anderson AT (1974) Flood hazard studies in the Mississippi River Basin using remote sensing. Water Resour Bull 10(5):1060–1081

Rango A, Solomon VV (1974) The utility of short wavelength remote sensing techniques for the monitoring and assessment of hydrological parameters. In: Proceedings of 11th international symposium on remote sensing of environment, Ann Arbor, MI, pp 55–64

Rashed T, Weeks J (2003a) Assessing vulnerability to earthquake hazards through spatial multicriteria analysis of urban areas. Int J Geogr Inf Sci 17(6):547–576

Rashed T, Weeks J (2003b) Exploring the spatial association between measures from satellite imagery and patterns of urban vulnerability to earthquake hazards. Int Arch Photogramm Remote Sens Spat Inf Sci XXXIV-7(W9):144–152

Rashed R, Weeks J, Couclelis H, Herold M (2007) An integrative GIS and remote sensing model for place-based urban vulnerability analysis. In: Mesev V (ed) Integration of GIS and remote sensing. Wiley, Chichester, pp 199–224

Rasid H, Pramanik MAH (1990) Visual interpretation of satellite imagery for monitoring floods in Bangladesh. Environ Manage 14(6):815–821

Rejeski D (1993) GIS and risk: a three-culture problem. In: Goodchild MF, Parks BO, Steyaert LT (eds) Environmental modelling with GIS. Oxford University Press, New York, pp 318–331

Richards JA, Woddgate PW, Skidmore AK (1987) An explanation of enhanced radar backscatter from flooded forests. Int J Remote Sens 8(7):1093–1100

Sadars R, Tabuchi S (2000) Decision support system for flood risk analysis for the river Thames, United Kingdom. Photogramm Eng Remote Sens 66(10):1185–1193

Sakamoto T, Nguyen NV, Kotera A, Ohno H, Ishitsuka N, Yokozawa M (2007) Detecting temporal changes in the extent of annual flooding within the Cambodia and the Vietnamese Mekong delta from MODIS time-series imagery. Remote Sens Environ 109(3):295–313

Sanyal J, Lu XX (2004) Application of remote sensing in flood management with special reference to Monsoon Asia: a review. Nat Hazard 33(2):283–301

Sanyal J, Lu XX (2005) Remote sensing and GIS-based flood vulnerability assessment of human settlements: a case study of Gangetic West Bengal, India. Hydrol Process 19(18):3699–3716

Sanyal J, Lu XX (2006) GIS based flood hazard mapping in Gangetic West Bengal. Singap J Trop Geogr 27(2):207–220

Sanyal J, Lu XX (2009) Ideal location for flood shelter: a geographic information system approach. J Flood Risk Assess 2(4):262–271

Scheuer S, Haase D, Meyer V (2011) Exploring multicriteria flood vulnerability by integrating economic, social and ecological dimensions of flood risk and coping capacity: from a starting point view towards an end point view of vulnerability. Nat Hazard 58(2):731–751

Schneiderbauer S, Ehrlich D (2004) Risk, hazard and people's vulnerability to natural hazards: a review of definitions, concepts and data. European Commission – Joint Research Centre (ERC-JRC), Brussels

Schumann G, Baldassarre GD (2010) The direct use of radar satellites for event-specific flood risk mapping. Remote Sens Lett 1(1):75–84

Sebastian S, Haase D, Meyer V (2011) Exploring multicriteria flood vulnerability by integrating economic, social and ecological dimensions of flood risk and coping capacity: from a starting point view towards an end point view of vulnerability. Nat Hazard 58(2):731–751

Sheng Y, Su Y, Xiao Q (1998) Challenging the cloud-contamination problem in flood monitoring with NOAA/AVHRR imagery. Photogramm Eng Remote Sens 64(3):191–198

Showalter PS (2001) Remote sensing's use in disaster research: a review. Disast Prev Manage 10(1):21–29

Showalter PS, Lu Y (eds) (2010) Geospatial techniques in urban hazards and disaster analysis. Geotechnologies and the environment, vol 2. Springer, Dordrecht

Simonovic SP (1993). Flood control management by integrating GIS with expert systems: Winnipeg city case study. In: HydroGIS'93: applications of geographic information systems in hydrology and water resources. IAHS Pub no. 211. International Association of Hydrological Sciences, Oxford

Sinha R, Bapalu GV, Singh LK, Rath B (2008) Flood risk analysis in the Kosi river basin, North Bihar using multi-parametric approach of analytical hierarchy process (AHP). J Indian Soc Remote Sens 36(4):293–307

Sirikulchayanon P, Sun W, Oyana TJ (2008) Assessing the impact of the 2004 tsunami on mangroves using remote sensing and GIS. Int J Remote Sens 29(12):3553–3576

Skidmore AK, Woodgate PW, Richards JA (1986) Classification of the Riviera forest of south eastern Australia using co-registered Landsat MSS and SIR-B radar data. In: Proceedings of the symposium on remote sensing for resources development and environmental management. A.A. Balkema, Rotterdam, pp 517–519

Smith DI (1981) Actual and potential flood damage: a case study for urban Lismore, NSW, Australia. Appl Geogr 1(1):31–39

Smith LC (1997) Satellite remote sensing of river inundated area, stage and discharge: a review. Hydrol Process 11:1427–1439

Smith K, Ward R (1998) Floods: physical processes and human impacts. Wiley, Chichester

Solberg SAH, Jin AK, Taxt T (1994) Multisource classification of remotely sensed data: fusion of Landsat TM and SAR images. IEEE Trans Geosci Remote Sens 32(4):768–778

Solbø S, Solheim I (2004) Towards operational flood mapping with satellite SAR. In: Envisat/ERS symposium, Salzburg

Solbø S, Pettinato S, Paloscia S, Santi E, Brusotti P, Solheim I (2004) Mapping of flooding in the Alessandria area with ERS. In: Proceedings of IGARSS 2004, Anchorage, Alaska, pp 4689–4692

Srivastava YK, Pal DK, Das RK, Sudhakar S, Adiga S, Venkatachary KV, Srivastava SK (2000) High resolution remote sensing data and GIS techniques in updating of infrastructures details for flood damage assessment – a case study. http://www.gisdevelopment.net/aars/acrs/2000/ts8/hami0002pf.htm. Accessed 2 Aug, 2004

Su M-D, Kang J-L, Chang L-F, Chen AS (2005) A grid-based GIS approach to regional flood damage assessment. J Marine Sci Technol 13(3):184–192
Suriya S, Mudgal BV (2012) Impact of urbanization on flooding: the Thirusoolam sub watershed – a case study. J Hydrol 412–413:210–219
Taft OW, Haig SM, Killsgaard C (2003) Use of radar remote sensor (RADARSAT) to map wetland habitat for shorebirds in an agricultural landscape. Environ Manage 32(2):268–281
Tanavud C, Yongchalermchai C, Bennui A, Densreeserekul O (2004) Assessment of flood risk in Hat Yai Municipality, Southern Thailand, using GIS. J Nat Disast Sci 26(1):1–14
Tate E, Burton CG, Berry M, Emrich CT, Cutter SL (2011) Integrated hazard mapping tool. Trans GIS 15(5):689–706
Taubenböck H, Post J, Roth A, Zosseder K, Strunz G, Dech S (2008) A conceptual vulnerability and risk framework as outline to identify capabilities of remote sensing. Nat Hazard Earth Syst Sci 8(3):409–420
Taubenböck H, Roth A, Dech S (2009) Megacities: hints for risk assessment using EO data. In: Gamba P, Herold M (eds) Global mapping of human settlement: experiences, datasets and prospects. CRC Press, Boca Raton, pp 205–230
Thomas DSK, Mitchell JT (2001) Which are the most hazardous states? In: Cutter SL (ed) American hazardscapes: the regionalization of hazards and disasters. Joseph Henry Press, Washington, DC, pp 115–156
Timmerman P (1981) Vulnerability, resilience and the collapse of society. Environmental monograph. Institute for Environmental Studies, University of Toronto, Toronto
Tingsanchali T, Karim MF (2005) Flood hazard and risk analysis in the southwest region of Bangladesh. Hydrol Process 19(10):2055–2069
Townsend PA (2001) Mapping seasonal flooding in forested wetlands using multi-temporal RADARSAT SAR. Photogramm Eng Remote Sens 67(7):857–864
Townsend PA, Walsh SJ (1998) Modeling floodplain inundation using an integrated GIS with radar and optical remote sensing. Geomorphology 21(3–4):295–312
Töyora J, Pietroniro A, Martz LW (2002) A multi-sensor approach to wetland flood monitoring. Hydrol Process 16:1569–1581
Tralli DM, Blom RG, Zlotnicki V, Donnellan A, Evans DL (2005) Satellite remote sensing of earthquake, volcano, flood, landslide and coastal inundation hazards. ISPRS J Photogramm Remote Sens 59(4):18198
Tran P, Shaw R, Chantry G, Norton J (2009) GIS and local knowledge in disaster management: a case study of flood risk mapping in Vietnam. Disasters 33(1):152–169
Turner BL, Kasperson RE, Matson PA, McCarthy JJ, Corell RW, Christensen L, Eckley N, Kasperson JX, Luers A, Martello ML, Polsky C, Pulsipher A, Schiller A (2003) A framework for vulnerability analysis in sustainability science. Proc Natl Acad Sci 1000(14):8074–8079
Uitto JI (1998) The geography of disaster vulnerability in megacities. Appl Geogr 18(1):7–16
Ulaby F, Moore R, Fung A (1981) Microwave remote sensing: active and passive, vol 2 microwave remote sensing fundamentals and radiometry. Artech House, Norwood
United Nations (UN) (2003) Guidelines for reducing flood losses. Available at http://www.un.org/esa/sustdev/publications/flood_guidelines.pdf. Accessed 12 Jan, 2009
United Nations (UN) (2005). Hyogo framework for action 2005–2015 – building the resilience of nations and communities to disasters. In: World conference on disaster reduction, Kobe, Japan, 18–22 January
United Nations Development Programme (UNDP) (1992) An overview of disaster management, 2nd edn. UNDP, Washington, DC
United Nations Environmental Program (UNEP) (2003) Assessing human vulnerability to environmental change: concepts, issues, methods and case studies. UNEP, Nairobi
UN/ISDR (International Strategy for Disaster Reduction) (2004) Living with risk: a global review of disaster reduction initiatives. United Nations Publications, Geneva
van der Sande CJ, de Jong SM, de Roo APJ (2003) A segmentation and classification of IKONOS-2 imagery for land cover mapping to assist flood risk and flood damage assessment. Int J Appl Earth Observ Geoinf 4(3):217–229

van Oosterom P, Zlatanova S, Fendel EM (2005) Geo-information for disaster management. Springer, Berlin
van Westen CJ (1993) Remote sensing and geographic information systems for geological hazard mitigation. ITC J 4:393–399
Wadge G, Wislocki AP, Pearson EJ (1993) Spatial analysis in GIS for natural hazard assessment. In: Goodchild MF, Parks BO, Steyaert LT (eds) GIS and environmental modelling. Oxford University Press, New York, pp 332–338
Waisurasingha C, Aniya M, Hirano A, Sommut W (2008) Use of Radarsat-1 data and a digital elevation model to assess flood damage and improve rice production in the lower part of the Chi river basin, Thailand. Int J Remote Sens 29(20):5837–5850
Waite WP, Macdonald HC, Kaupp VH, Demarcke JS (1981) Wetland mapping with imaging radar. In: International geoscience and remote sensing symposium, Washington DC, 8–10 June, pp 794–799
Wang Y (2004) Using Landsat 7 TM data acquired days after a flood event to delineate the maximum flood extent on a coastal floodplain. Int J Remote Sens 25(5):959–974
Wang J-F, Li L-F (2008) Improving tsunami warning systems with remote sensing and geographical information systems input. Risk Anal 28(6):1653–1668
Wang Y, Hess LL, Filso S, Melack JM (1995) Understanding the radar backscattering from flooded and non-flooded Amazonian forest: results from canopy backscatter modeling. Remote Sens Environ 54(3):324–332
Wang Y, Colby JD, Mulcahy KA (2002) An efficient method for mapping flood extent in a coastal floodplain using Landsat TM and DEM data. Int J Remote Sens 23(18):3681–3696
Wang Y, Li Z, Tang Z, Zeng G (2011) A GIS-based spatial multi-criteria approach for flood risk assessment in Dongting lake region, Hunan, Central China. Water Resour Manage 25(13):3465–3484
Watson CC Jr (1992) GIS aids hurricane planning. GIS World, July (Special Issue):46–52
Watts MJ, Bohle HG (1993) The space of vulnerability: the causal structure of hunger and famine. Prog Hum Geogr 17(1):43–67
Webster TL, Forbes DL, Dickie S, Shreenan R (2004) Using topographic lidar to map flood risk from storm-surge events for Charlottetown, Prince Edward Island, Canada. Can J Remote Sens 30(1):64–76
Webster TL, Forbes DL, MacKinnon E, Roberts D (2006) Flood-risk mapping for storm-surge events and sea level rise using lidar for southeast New Brunswick. Can J Remote Sens 32(2):194–211
Weichselgartner J (2001) Disaster mitigation: the concept of vulnerability revisited. Disast Prev Manage 10(2):85–94
Weng Q (2001) Modeling urban growth effects on surface runoff with the integration of remote sensing and GIS. Environ Manage 28(6):737–748
Werner MGF (2001) Impact of grid size in GIS based flood extent mapping using a 1D flow model. Phys Chem Earth Part B: Hydrol Oceans Atmos 26(7–8):517–522
White GF (ed) (1974) Natural hazards: local, national, global. Oxford University Press, New York
White GF, Haas JE (1975) Assessment of research on natural hazards. MIT Press, Cambridge
White GF, Kates RW, Burton I (2001) Knowing better and losing even more: the use of knowledge in hazards management. Environ Hazard 3(3–4):81–92
Wiesnet DR, McGinnis DF, Pritchard JA (1974) Mapping of the 1973 Mississippi River floods by NOAA-2 satellite. Water Resour Bull 10(5):1040–1049
Willett K, Sharda R (1991) Using the analytical hierarchy process in water resources planning: selection of flood control projects. Socioecon Plann Sci 25(2):103–112
Williamson AN (1974) Mississippi river flood maps from ERTS-1 digital data. Water Resour Bull 10(5):1050–1059
Wisner B (1993) Disaster vulnerability: scale, power and daily life. GeoJournal 30(2):127–140
Wisner B, Blaikie P, Cannon T, Davis I (2004) At risk: natural hazard, people's vulnerability and disasters, 2nd edn. Routledge, Abingdon

Wu SY, Yarnal B, Fisher A (2002) Vulnerability of coastal communities to sea-level rise: a case study of Cape May County, New Jersey, USA. Clim Res 22:255–270

Yahaya S (2008) Multicriteria analysis for flood vulnerable areas in Hadejia-Jama' Are river basin, Nigeria. ASPRS annual conference, Oregon, April 28–May 2

Yalcin G, Akyurek Z (2004) Analysing flood vulnerable areas with multicriteria evaluation. Available at http://www.isprs.org/proceedings/XXXV/congress/comm2/papers/154.pdf. Accessed 15 Jan, 2012

Yamagata Y, Akiyama T (1988) Flood damage analysis using multitemporal Landsat thematic mapper data. Int J Remote Sens 9(3):503–514

Yoon DK (2012) Assessment of social vulnerability to natural disasters: a comparative study. Nat Hazard 63(2):823–843

Zahran S, Brody SD, Peacock WG, Vedlitz A, Grover H (2008) Social vulnerability and the natural and built environment: a model of flood causalities in Texas. Disasters 32(4):537–560

Zerger A (2002) Examining GIS decision utility for natural hazard risk modelling. Environ Model Software 17(3):287–294

Zerger A, Smith DI (2003) Impediments to using GIS for real-time disaster decision support. Comput Environ Urban Syst 27(2):123–141

Zerger A, Wealands S (2004) Beyond modelling: linking models with GIS for flood risk management. Nat Hazard 33(2):191–208

Zhang J, Hori T, Tatano H, Okada N, Matsumoto T (2003) GIS and flood inundation model-based flood risk assessment in urbanized floodplain. In: Cheng Y et al (eds) GIS and RS in hydrology, water resources and environment, vol 1. Sun Yat-sen University Press, Guangzhou, pp 92–99

Zheng N, Takara K, Yamashiki Y, Tachikawa Y (2009) Large scale quantitative vulnerability analysis for regional flood hazards. Annuals of Disaster Prevention and Research Institute, Kyoto University, No. 52B, Japan

Zhou YC, Qing CW (1999) Deciding the flood extent with RADARSAT SAR data and image fusion. http://www.gisdevelopment.net/aars/acrs/1999/ps3/ps3999.shtml. Accessed 11 July, 2003

Zhou C, Luo J, Yang C, Li B, Wang S (2000) Flood monitoring using multi-temporal AVHRR and RADARSAT imagery. Photogramm Eng Remote Sens 66(5):633–638

Chapter 3
Vulnerability of a Megacity to Flood: A Case Study of Dhaka

Abstract This chapter illustrates the physical environment together with the built and socioeconomic characteristics of the megacity of Dhaka. An account of historical floods with damage statistics is discussed, revealing that flood loss in the city has increased over time. This is clearly attributed to unplanned and ill-structured development planning. A number of factors are accountable for the increase in flood vulnerability, including extreme population density, intense inequality in resource distribution, and dilapidated drainage systems. Although various adaptation strategies have been suggested for flood management, their success depends on a number of issues such as restricting urban expansion in flood-prone areas.

3.1 Introduction

Dhaka, the capital of the People's Republic of Bangladesh, first attracted attention when it became the provincial capital in 1905 (Ahmed 1986). After the partition of the Indian subcontinent in 1947 into India and Pakistan, Dhaka became the capital of the then East Pakistan. Bangladesh emerged as an independent state in 1971, and the country retains Dhaka as its capital. Since then, Dhaka has been the major focus of administrative, social, educational, and cultural activities. While the city is subject to regular inundation during the monsoon season, it is only relatively recently that flooding has become a matter of serious apprehension. Due to its rapid development, which is primarily driven by population growth, flood becomes pervasive in the wet season, causing colossal damage to property and severely disrupting the regular functioning of the urban system.

Section 3.2 describes Dhaka's physical environment, and Sect. 3.3 explains the growth of Dhaka along with its present built and socio-environmental conditions. A historical account of floods and the causes of Dhaka's increasing vulnerability to flood is illustrated in Sect. 3.4.

A.M. Dewan, *Floods in a Megacity: Geospatial Techniques in Assessing Hazards, Risk and Vulnerability*, Springer Geography, DOI 10.1007/978-94-007-5875-9_3,

3.2 Physical Environment

The establishment of the municipality in Dhaka can be traced back to 1830. At that time, it was known as the "Dhaka Committee," and it was renamed to Dhaka Municipality in 1864 (Rizvi 1969). After independence, Dhaka Municipality was replaced with Dhaka Paurashava under the Bangladesh Local Council and the Municipal Order 1972, which was later upgraded to Dhaka City Corporation (DCC) in 1991 (Banglapedia 2006). Dhaka attained the status of "megacity" in 1991 with a population of 6.8 million (BBS 2008). The total area of the Dhaka megacity is 1,383 km^2, and it comprises of the DCC, five adjacent municipalities (Tongi, Savar, Gazipur, Narayanganj, and Kadamrasul), and a number of urban areas of which the DCC has an area of 145 km^2 (Islam 2005a).

There are five major rivers flowing across Dhaka, namely, the Buriganga to the south, Turag to the west, Tongi Khal to the north, and the Lakhya and Balu rivers to the east. Dhaka and its adjoining areas are composed of alluvial terraces of the southern part of the Madhupur tract and low-lying areas at the doab of the river Meghna and Lakhya. In course of time, this tract has been merged and dissected by the recent floodplains on its fringe to form the present landform of Dhaka. The major geomorphic units of the area are the highlands (or the Dhaka Terrace), the lowlands or floodplains, depressions, and abandoned channels (Miah and Bazlee 1968). Geomorphic classification reveals that a relatively young floodplain constitutes the largest area, followed by the higher terraces of the Pleistocene period (Kamal and Midorikawa 2004). Low-lying swamps and marshes located in and around the city are examples of other major geomorphic features. The elevation of Dhaka ranges from 1 to 16 m above the mean sea level (FAP 8A 1991).

The climate of Dhaka can be classified as tropical monsoon and is characterized by three distinct seasons: summer, winter, and rainy season. The average annual rainfall is about 2,000 mm, of which 80% occurs during the monsoon season. The temperature during the summer months ranges between 28 and 34°C. In winter, the temperature ranges between 10 and 21°C.

3.3 Built and Socioeconomic Environments

The growth of Dhaka can be categorized into five unique periods: the pre-Mughal period, the Mughal period, the British colonial period, the Pakistan period, and the Bangladesh period. The growth of Dhaka saw many ups and downs during the first four periods (Islam 1996). From 1971, the growth of Dhaka has become phenomenal (Chowdhury and Faruqui 1991). Although there is some confusion about the exact number of urban areas, a recent study revealed that Dhaka is expanding rapidly following the independence of the country (Ahmed et al. 2012).

Historically, Dhaka had an excellent natural drainage system comprising intricate river networks with numerous *khals* (ephemeral water bodies) and canals that used to drain water from its upper reaches during monsoons. At least 40 large *khals*,

Table 3.1 Demographic characteristics of Dhaka megacity

Year	Total households	Population	Density (per km^2)	Sex ratio	Literacy rate (%)	Household size (average)	Growth rate (%)
1961	127,710	718,766	5,796	154	–	5.6	–
1974	341,167	2,068,353	6,156	137	–	6.1	11.15
1981	527,311	3,440,147	8,547	139	57.0	6.0	5.22
1991	1,088,378	6,487,459	4,795	126	62.3	5.4	6.55
2001	1,920,682	9,672,763	7,055	125	65.1	4.6	4.08
2011	3,232,683	14,509,100	10,484	113	67.3	4.1	–

Source: BBS (1991, 2008, 2012)

numerous ponds, creeks, and five major rivers used to drain 80% of the city's floodwater during the wet season (Khan 2000, 2006).

The city has extensive administrative and infrastructure facilities, as well as extensive road and telecommunication networks. Hence, it becomes the focus of urban expansion and the hub of economic activities. Due to its economic and sociopolitical significance, marginalized rural people are attracted to the area for better employment opportunities and improved lifestyles. As a result, Dhaka has become one of the fastest-growing cities in the world, primarily driven by explosive population growth. The city's population was 0.41 million in 1951 and 0.71 million in 1961. By 1974, it had risen to 2.06 million, averaging an annual growth rate of 11.15% between 1961 and 1974 (BBS 2008; Islam 1996, 1999, 2005a). In 1981, the population rose to 3.44 million. The population reached around 6.48 million in 1991 and 9.6 million in 2001 (BBS 2001, 2003) (see Table 3.1). Currently, Dhaka's population is more than 14 million, with an average annual growth rate of 4.08% during 1991–2001, which outpaced the country's annual growth rate of 1.3% (BBS 2012, 2011; World Bank 2007). Dhaka was one of the top 10 megacities in the world in 2011; if the current rate continues, Dhaka will replace Beijing by 2025, with a projected population of 22.9 million (UN 2012).

As seen in Table 3.1, the average household size and the sex ratio of Dhaka is declining, indicating increases of single-family households and the number of single females in the urban workforce (Siddiqui et al. 1993). Studies indicate that the rapid growth of the urban population is mainly driven by rural–urban migration. Islam (1991) reported that more than 60% of people in Dhaka are migrants. A recent study showed that Dhaka receives 300,000–400,000 migrants every year (Sanderson 2012). The observed age-specific population supports this estimation; the number of young people is higher compared with other age groups, suggesting that rural–urban migration is age selective (Afsar 2000). In addition to rural–urban migration, another explanation exists for the phenomenal growth of Dhaka after the independence of Bangladesh. For example, Dutt and Noble (2004) noted that, prior to independence, urban growth in Bangladesh was very slow owing to the fact that most of the investment during that time took place in Pakistan or elsewhere and Bangladesh was put in the backwaters, which may have resulted in lower urban expansion before independence. The sovereignty in 1971 allowed the country to propagate Dhaka as the heart of economic and social development, thereby providing better job opportunities, wages, infrastructure, and other public services, which encouraged rural people to migrate. Understandably, these migrants exert tremendous pressure on land for housing and other urban services, which in turn has profound environmental implications, including flood.

Extreme inequality exists between the rich and the poor in Dhaka (Begum 2007; Islam 2005c), as indicated by the lowest per capita income of US$327 (CUS 1990). This is perhaps the lowest per capita income of all megacities around the world (Siddiqui et al. 2004). Rich people constitute only 3% of the total population who enjoy a high standard of living, while the rest are in the middle- and lower-income groups (Hossain 2006). Consequently, a significant amount of the population lives below the poverty line and struggles to survive. In addition, 34.3% of the total population is in the 0–14 and 60+ age groups, suggesting a high-dependency burden on the working age group (BBS 2008). The literacy rate is rising steadily, and it is higher among males. According to the 2001 population census, the male literacy rate was 69.8%, while the female literacy rate was 59.2% (BBS 2003).

Sources of drinking water in Dhaka vary according to the locality, although groundwater remains the major source (Ahmed et al. 2005). Fifty-three percent of households have access to piped water, while 43.1% rely on tube wells for their potable water sources (BBS 2008). The rest of the households depend on deep tube wells, ponds, and other sources such as rivers. As the population is growing rapidly, it has become difficult for the Dhaka Water Supply and Sewerage Authority (DWASA) to ensure pipe water for every household. Estimates show that DWASA is currently facing a water shortage of about 500 million liters per day, and the amount is likely to increase by 2015 (Islam et al. 2011). Among inhabitants in Dhaka, slum dwellers face an acute shortage of potable water (Akbar et al. 2007). The sanitation system in Dhaka is also in a chronic crisis. It is estimated that 15–20% of people are being served by the sewage system, 25% have a septic tank on site, and around 35–40% of people do not have access to safe sanitation (Hossain 2006; Haq 2006); therefore, these people rely on temporary latrines made from bamboo and straw. Further, it is estimated that over 70% of slum dwellers do not have access to safe sanitation (CUS et al. 2006).

Due to increasing anthropogenic activities, the quality of surface water has been deteriorating in the peripheral rivers and other water bodies. As Dhaka hosts more than 40% of the country's industries (BBS 2005), industrial effluents are directly discharged into the rivers, and are the primary cause of water contamination (Hossain and Rahman 2012; Dewan et al. 2012a; BBS 2010; Rahman and Hossain 2008; Sohel et al. 2003; Karn and Harada 2001; Kamal et al. 1999). Slum dwellers typically build open latrines on the roadside or construct hanging latrines on the water bodies. When natural and storm water runoff transport human waste to the surrounding water bodies, it results in severe water pollution (Dewan et al. 2012a; BBS 2010; Hossain 2006; Hasan and Mulamoottil 1994). In addition, the quality of piped water and groundwater is questionable too (Aktar et al. 2009; Varis et al. 2006). For example, the zinc concentration in Dhaka's water supply is much higher than the admissible limit (Maroof et al. 1986). Groundwater quality is also decreasing, particularly in the southern part, where leather industries discharge untreated effluents into inland water bodies (Zahid et al. 2006; Karn and Harada 2001).

It is important to note that the area under current investigation is a subset of Dhaka Metropolitan Development Plan (DMDP) area and includes the entire DCC, Tongi, and Savar municipalities with adjacent *unions* and other urban areas. It is located between 23.61° and 23.97°N and 90.22° and 90.59°E, and it covers an area of 878 km^2 (see Fig. 3.1). Due to the unavailability of data (Huq 1999), the study had to be restricted to this spatial extent.

3.4 Floods in Dhaka: History and Causes of Vulnerability

Similar to other parts of Bangladesh, flooding by spilling of river waters is a frequently occurring phenomenon in Dhaka (FAP 8A 1991). Although fluvial flooding is the most costly and debilitating natural hazard and remains one of the greatest threats of all natural hazards to people and property, pluvial flooding is becoming a grave concern for the inhabitants of Dhaka (Stalenberg and Vrijling 2009; Alam and Rabbani 2007; Tawhid 2004). While Dhaka's peculiar geographical location makes it particularly vulnerable to flood (Bromley et al. 1989), piecemeal and uncoordinated urbanization has accelerated the degree of vulnerability to flood, particularly in the recent past (Dewan and Yamaguchi 2008; Islam 2005b).

Dhaka has experienced many devastating floods in the past. Historical accounts show that it was heavily inundated in 1787 and 1788 following by the heavy monsoonal downpour, and the depth of flooding was adequate enough to admit boats sailing on the streets in the nucleus of the city. In 1833–1834 and 1870, a number of floods inflicted Dhaka and its adjacent areas (Hunter 1877). Floods that occurred in the 1950s and 1960s also sternly affected the city (Rizvi 1969). Contemporary flood history revealed that the lowlands of Dhaka have been regularly inundated during monsoons since 1954; however, in many instances, the flood situation depends on the amount of precipitation and flows from the upstream points (Faisal et al. 1999, 2003; Nishat et al. 2000). Disastrous floods in 1987 and 1988 inundated areas of 164 and 200 km^2, respectively (FAP 8A 1991). The unprecedented flood of 1998 also severely affected Dhaka and its neighboring areas (DMB 1998), which resulted in unusual damage and myriad sufferings to the people. In total, the 1988 flood affected 4.55 million people (Hye 2000), and most of Dhaka was under water to various depths for more than 8 weeks (Jahan 2000). Dhaka was again flooded in 2004 and 2007; however, the magnitude was not as destructive as the 1998 flood. Intense rainfall during September 2004 (e.g., 341 mm in 48 h) made the flood situation particularly worse (Hasnat 2004). Even though the embankment acted as a buffer to save lives and property in the 2004 and 2007 floods, pluvial flooding by means of localized ponding and water logging were seriously pronounced, particularly in the embanked part, which resulted in a number of environmental problems, including scarcity of drinking water, disruptions to economic activities, and the prevalence of water-borne diseases (Bala et al. 2009; Alam and Rabbani 2007; Khan 2006; Khan 2001; Tawhid 2004).

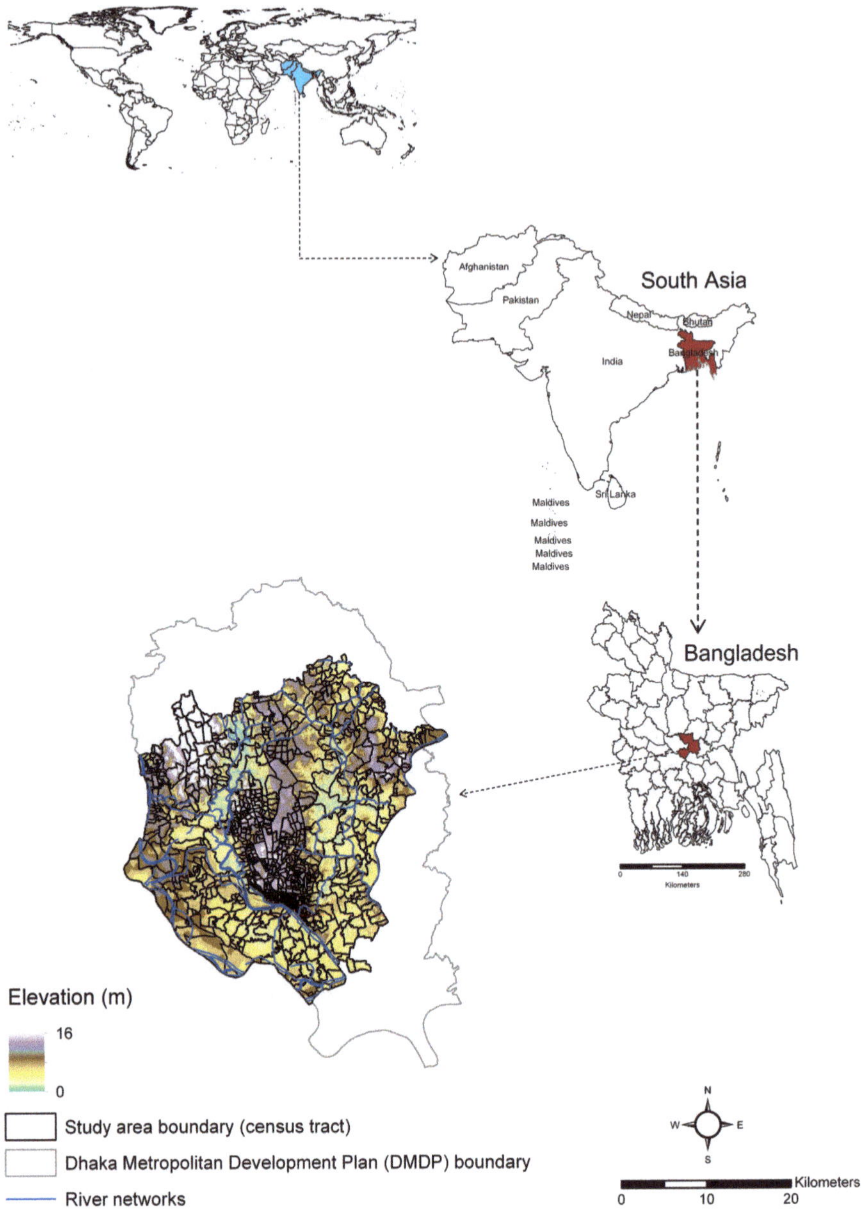

Fig. 3.1 Location of study area

In addition, the regular functioning of the urban system was severely disrupted due to water logging during the floods of 2004 and 2007.

Although flood is a regular feature, comprehensive damage reports are not available. It may be because different organizations estimate damage from different perspectives; therefore, consistent damage estimations are not available, or if they are available, damaged records tend to be incomplete, inconsistent, and overestimated (Chowdhury and Sato 2000). Nonetheless, this study attempted to document the damage statistics of major events by screening relevant literature. Among the four recent floods that shattered Dhaka, damage caused by the 1998 flood was the most severe, as the extent and scale of the flood excelled all previous records in terms of severity.

The hardest-hit sector was housing; nearly 262,000 houses of various types were damaged during the 1998 flood, worth $46.6 million (Taka 2.3 billion[1]) (Islam 1998). Around 1,000 km of various types of roads were damaged (Siddique and Chowdhury 1998). The estimated loss to the water supply and sanitation sector was $10 million, while massive damage was also reported in the power sector (Faisal et al. 1999). A total of 223 km of box culverts, open drains, and storm sewers were reported to have silted up or been damaged (Siddique and Chowdhury 1998). Large- and small-to-medium-scale industries in Dhaka were severely affected, and the total loss in these two sectors was estimated at around $30 and $36 million, respectively (Nishat et al. 2000). In the DCC area, 71 wards out of 90 were badly damaged, and about 3.7 million people were affected (Ali et al. 2002). Damage estimation with a questionnaire survey revealed that poor-quality residences and household goods such as furniture suffered significant damage (Chowdhury and Sato 2000). The total loss incurred by the 1998 flood was estimated to about US$3,000 million (Jahan 2000).

Available statistics on the 1988 flood revealed that the death toll was at least 150, and more than 2.2 million people were affected (Huq and Alam 2003). Eighty-five percent of the urban areas were engulfed with floodwaters for several weeks, and the depth of flooding ranged from 0.3 to 4.5 m (Johnson 2010). The number of institutions and houses affected by the 1988 flood was estimated at 14,000 and 400,000, respectively (Alam and Rabbani 2007). The total loss for residential buildings was $125 million (4 billion TK), and more than $12.5 million (400 million TK) was reported for institutional damage (Nishat et al. 2000).

Although fluvial flooding did not engulf most of the embanked areas of Dhaka, losses due to pluvial flooding were appalling in 2004 (Ahmed et al. 2006). For instance, infrastructural damage in terms of road repair costs was estimated at $208.2 million (12.8 billion TK), and the loss in the telephone sector amounted to $2.8 million (175 million TK). The ready-made-garments sector experienced a loss of $10.3 billion (632 billion TK) (Alam and Rabbani 2007), illustrating the degree of loss due to pluvial floods. An assessment of flood damage in terms of urban areas under water indicated that the 1988 flood engulfed 1,484 hectares (ha) of urban land, whereas the figure was 2,991 ha in 1998. It is interesting to note that the moderate flood of 2004 (compared to the 1998 flood) engulfed 4,503 ha of urban land (Dewan and Yamaguchi 2007, 2008),

---

[1] One US$ = 32 TK in 1988, 49.3 TK in 1998 and 61.5 TK in 2004.

which clearly indicates that the extent and damage of floods is on the rise because of the progressive construction of additional infrastructures in flood-prone areas.

In addition to damage to property and human fatalities, a secondary effect of flood is now creating a hue and cry among urban dwellers in Dhaka: the spread of water-borne diseases such as diarrhea, dysentery, and typhoid. For instance, Kafiluddin (2000) reported that 191,967 people with different water-borne diseases were admitted to different hospitals during the 1998 flood of which 284 people died. Similarly, cases of diarrhea and dysentery were higher during the 2004 flood (IEDCR 2004; Rahman et al. 2004). A total of 43,250 diarrhea cases were admitted to the ICDDR,B hospital during the 2007 flood (Harris et al. 2008). These studies demonstrate that the incidences of water-borne disease are rising with time, which has serious implications for public health in Dhaka.

What is the underlying factors contributing to Dhaka's flood vulnerability? A simple and straightforward answer is not possible, as urban vulnerability is the interaction of biophysical and socioeconomic factors, which is a complex and dynamic phenomenon (Lall and Deichmann 2012; Bhattarai and Conway 2010; Meyer et al. 2009; Taubenböck et al. 2008; Rashed et al. 2007; Rashed and Week 2003; Godschalk 2003; Cutter et al. 2000; Mitchell 1999a, b). In case of Dhaka, flood vulnerability is also the interplay of physical and human factors. Although it is difficult to identify the exact causes of vulnerability in the occasion of uncontrolled development triggered by rapid population growth, recognizing the causal factors that are accelerating vulnerability on a megacity scale would allow future risks to be minimized.

3.4.1 Low Elevation

As mentioned above, the ground elevation of Dhaka megacity is very low. Except for the Pleistocene terraces on the northern edge and further north, 71.5% of the land is 1–5 m above mean sea level (MSL). Around 26.7% of the land is 6–10 m, and the rest of the land (1.4%) is 11–16 m above the mean sea level. Most of the built-up areas are located 5–8 m above sea level. Due to its physical geography, floods triggered by monsoonal downpours in conjunction with the upstream flow bring death and injury to people and result in substantial damage to buildings and infrastructures.

3.4.2 Unplanned Urbanization

Unplanned and haphazard urban expansion is one of the most important factors intensifying flood hazards in many megacities around the world, including Dhaka (Benson and Clay 2003). Using a topographic map from 1960 and a series of Landsat data from 1975 to 2011, Dewan and Yamaguchi (2009a, b), Dewan et al. (2012b) and Dewan and Corner (2012) mapped out the spatial patterns of urban expansion in Dhaka. Griffiths et al. (2010) also examined patterns of urban growth between 1990 and 2006 using multi-sensor data. These

Plate 3.1 Floodplain development along the embanked part of Dhaka (Source: author)

studies demonstrated that urban growth is extremely faster, leading to a considerable increase of impervious surfaces by converting natural land such as productive agricultural land and precious wetlands. For example, the built-up areas increased from 4,625 ha in 1960 to 14,486 ha in 1999. In contrast, wetlands decreased from 13,514 ha to 10,797 ha during the same period (Dewan and Yamaguchi 2009a, b). Current estimates show that the total urbanized area reached about 19,407 ha in 2011, while natural land-use categories decreased exponentially (Dewan and Corner 2012). In 1961, inland water bodies constituted 5.1% of the entire DCC area, which decreased to about 1.8% by 2001 (Sultana et al. 2009; Islam 2009), indicating the degree of human encroachment on natural land. Studies reveal that the construction of the embankment in the western limb of Dhaka has accelerated the development of low-lying areas in the recent past (Ahmed et al. 2012; Ahmed and Ahmed 2012; Islam and Ahmed 2011; Alam and Ahmad 2010; Chowdhury and Sato 2000; Maathuis et al. 1999; Chowdhury et al. 1998) (Plate 3.1). As Pelling (2003, p. 49–50) noted:

> *'The presence of physical hazard defences such as flood-walls or landslide-retaining walls have two effects on the social geography of the city. First, these investments encourage formal or informal development on hazardous sites. Whilst engineering responses provide much protection, there are many examples of cases where hazards have surpassed the threshold of safety provided by physical structures with very high human costs. [....] Second, the presence of environmental infrastructure can increase the market value of protected land, which can result in low-income groups being replaced by middle- and high-income groups. This is most often the case when hazardous sites are located on desirable real estate land close to the city centre [...] whilst the physical component of hazard analysis is undoubtedly critical, this approach underestimates the importance of socio-economic or political forces in shaping the production and distribution of risks and vulnerability'.*

Approximately 6,000 ha of land for the Dhaka–Narayangonj–Demra (DND) project was originally retained to assist agricultural production in the 1960s. Locals and migrants have used this project area for residential purposes since the 1990s without approval from the relevant authority (Dewan and Yamaguchi 2009a; Islam 1996). This trend has continued throughout the megacity, for example, encroachment on low-lying lands in the Balu River catchment became evident after 2007, and it will have serious implications for future floods (Masood and Takeuchi 2012; Haque et al. 2012). This type of urban growth is identified as "uncoordinated, haphazard and ill-structured" (Islam 1996, 2005a), and it resulted in a significant increase of surface runoff (Chowdhury et al. 1998). Consequently, the responsible authority is unable to adequately plan and control development, which is gradually diminishing Dhaka's ability to deal with flood hazards.

While rapid urban growth has resulted in a considerable increase in flood extent and damage (Mohit and Akhter 2002), the management of floods and floodplains is a neglected priority for urban planning. From a hydrological point of view, floods that occurred in Dhaka in the 1970s were not significantly different from those that have occurred recently; however, flood duration has been prolonged due to rapid changes in land use and cover (Dewan and Yamaguchi 2008). The degree of economic loss and the vulnerability of populations, particularly in Dhaka, is thus the outcome of increased human occupancy in floodplains.

3.4.3 Dilapidated Drainage Systems

Since urban expansion increases imperviousness, an improved drainage system is a prerequisite to ensure the conveyance of floodwaters during monsoons. However, the drainage system in Dhaka has not been evolved in pace with the rapid urban development. Historically, Dhaka used to have excellent natural-gravity drainage comprising a number of lakes, canals, creeks, ponds, and *khals*. However, many of them are no longer exist (Khan 2006; Karim et al. 2005); for example, of 40 large *khals*, 17 are completely lost (Alam and Rabbani 2007). Some of the remaining *khals* lost their connectivity to development and illegal encroachment (Hossain and Rahman 2011). Currently, 225 km of subsurface pipes and 8 km of box culverts comprise the drainage system that covers Dhaka's 12 drainage zones (Okubo et al. 2010). In addition, a pumped drainage facility is in place (Khan 2008), but it is unable to function consistently during monsoons because of electricity shortages and pitiful maintenance (Faisal et al. 1993, 2003). Understandably, these drainage networks are inadequate for conveying huge volumes of floodwater generated by intense rainfall and fluvial systems during the wet season. The situation is further aggravated by the disposal and littering of solid wastes into the drainage system, which frequently causes blockages (Plate 3.2) in the decaying drainage infrastructure (Hossain and Rahman 2011). Inadequate drainage and dysfunctional storm water facilities unquestionably elevate Dhaka's flood damage potential.

Plate 3.2 Littering garbage and waste to the surface water bodies (Source: author)

3.4.4 Poor Performance of Flood Control Works

The government of Bangladesh, in association with development partners, has invested a significant amount of resources to protect its capital and adjoining areas from the perennial flood problem. Notable examples include the construction of embankments, floodwalls, and dykes. With respect to these efforts, there has been little success to date. The first flood-protection embankment in Dhaka was constructed in 1864 along the river Buriganga to secure the city against erosion and flood, and it is known as the Buckland Bund (Rizvi 1969). In the 1960s, there was another flood-protection project (referred to as DND), which was primarily aimed at saving cultivated lands from floods (Choudhury 2001). It was after the 1988 flood, when massive flood control works took place under the Flood Action Plan (FAP) in Bangladesh in general and Dhaka in particular. The objective of the FAP was to ensure the protection of the city against a once in 500–1,000-year fluvial flood and to minimize pluvial floods by using existing lakes, canals, and sewer systems (Stalenberg and Vrijling 2009). A range of other measures, such as sluice gates and

Plate 3.3 Effort of repairing a flood control structure during 1998 flood in eastern Dhaka (Source: Md. Mohiuddin Ahmed)

around 2,460 km of surface drains, are also in place (Barua and van Ast 2011). Although these flood control works provided safety to some parts of Dhaka during the recent floods, the overall performance of these flood control structures was poor for a number of reasons (Bala et al. 2009; Chowdhury et al. 1998; Miah and Mohit 1996). Lack of proper management due to fund constraint was found to be the key of them (Chowdhury et al. 1998). As Barua and van Ast (2011, p. 709) noted:

> *'If the flood prevention measures could have gained more preferences, the road maintenance costs would have been relatively less. But we do the opposite. We always look for short-term solutions. Long-term visioning requires institutional strengthening and substantial financial strength. In our economic system financial allocation for maintenance purpose does not exist. The ministry of Finance and Planning Commission makes financial arrangements only for implementation of the project work. But subsequent maintenance costs are not allocated for those projects.'*

Moreover, illegal encroachment, breaching, hydraulic leakage, waste disposal, and poor construction materials are other factors that contribute to the poor performance of flood control works in Dhaka (Stalenberg and Vrijling 2009; Faisal et al. 1993, 2003; Khan 2001; Chowdhury et al. 1998) (Plate 3.3).

Similar to other parts of Bangladesh, physical structures in Dhaka have also resulted in burgeoning environmental degradations, many of which are irremediable. For example, the flood extent has dramatically increased beyond the embanked areas as a result of changes in the hydrological regime linked with the embankment, causing significant agricultural damage and affecting a large number of the rural population (Dewan 2006). Rasid and Mallik (1996) identified several environmental problems associated with embankments. Pluvial floods have become widespread in response to the inconsiderate filling of surface water bodies along the embanked part (Stalenberg and Vrijling 2009), which poses a serious threat to social and economic activities (Alam and Rabbani 2007) (Plates 3.4–3.6 and 3.7). The water quality of the surface water bodies has severely degraded (Rana 2010; Alam et al. 1996; Rahman and Karim 1995), and a dramatic change in the quality of the living environment has also been reported (Chowdhury 2003). The poor performance of physical structures along with the environmental degradation associated with engineering measures is perpetuating the vulnerability of humans and infrastructures to floods.

3.4.5 High Population Density and Extreme Poverty

In addition to high population growth, population density is extremely high in Dhaka, making it one of the most crowded cities in the world (see Table 3.1). Due to the high rate of urbanization and the scarcity of land, people often construct their shelters in hazardous areas, for example, along riverbanks, on water bodies, and in the floodplains. As a result, spatial planning for emergency management has become a great hurdle.

As the growth of Dhaka is not commensurate with its economic development, the large rural–urban migration, particularly by marginalized rural people, has resulted in a rapid increase in the number of slum dwellers (Rana 2010; Angeles et al. 2009; Chowdhury and Amin 2006; Islam 2005c) (Plate 3.8). Although the incidences of poverty have recently been reduced (Mirza 2011), the percentage of urban poor is still as high as 45%, of which 25% is classified as extremely poor (Diaspora 2007 in Sanderson 2012). Table 3.2 shows that the slum population in Dhaka has more than doubled in the past decade, reaching 3.4 million in 2005. The number of slum clusters (nearly 5,000 at present) increased to about 70% during the same period (CUS et al. 2006), and the density of the slums is much higher than non-slum urban areas (Sanderson 2012). The physical environments of the slum settlements are very unhygienic (Plate 3.9), as they are often located in both unconventional and undesirable places, such as near solid waste dumps, open drains and sewers, in low-lying lands, on embankments, and along railway tracts (UN-HABITAT 2003). Consequently, the people living in these slums are extremely vulnerable to natural calamities, particularly floods (Braun and Aßheuer 2011; Jabeen et al. 2010; Rashid et al. 2007; Rashid 2000), because many of the slum dwellers live in close proximity to rivers and water bodies (World Bank Office 2007 in Stalenberg and Vrijling 2009).

Extreme population density, rampant poverty, and scarcity of resources are contributing to deplorable conditions, which have a profound effect on flood vulnerability. For instance, dense patterns of urban living can contribute to the generation of risk, and crowdedness amplifies the likelihood of disease transmission (Hewitt 1997). As noted by Stalenberg and Vrijling (2009, p. 476):

Plates 3.4–3.6 Sufferings of urban dwellers due to pluvial flooding in 2012 (Source: Md. Asaduzzaman Pramanik)

Plate 3.7 Floodwater engulfed Bangladesh Secretariat at the center of Dhaka (Source: Md. Asaduzzaman Pramanik)

Plate 3.8 Informal settlements in Dhaka (Source: author)

Table 3.2 Slum clusters and slum population in Dhaka, 1974–2005

	Year			
	1974	1991	1996	2005
No. of slum clusters	500	2,156	3,007	4,966
Total slum population (m)	0.27	0.7	1.5	3.4
Proportion of total population (%)	10	20	22	37.4

Source: CUS et al. (2006), Islam (1991, 1996), and Islam and Mahbub (1988)

Plate 3.9 Environmental quality of a slum (Source: author)

> *'Psychologically, a large part of the population of Dhaka is still on the lowest level of Maslow's hierarchy of needs, namely in the layer of physiological needs. Maslow states that if all needs are unsatisfied, and the organism is then dominated by physiological needs, all other needs may become simply non-existent or be pushed into the background'.*

Consequently, people in Dhaka, particularly the marginal groups, are more concerned about their existence than about the potential threat of natural hazards.

3.4.6 Flood Preparedness

As a nonstructural measure, flood forecasting and warning systems have been improved significantly under the FAP, which allows vulnerable communities to prepare well ahead of an impending disaster. It was found that the quality of services provided by

Plate 3.10 Vulnerability of marginal group to flooding (Source: Md. Mohiuddin Ahmed)

the flood forecasting and warning center (FFWC) of the Bangladesh Water Development Board (BWDB) was positive, but technical expertise, especially in estimating rainfall and flow to boundary conditions, appeared to be lacking for competent forecasting (Chowdhury 2000). In addition, warning messages in their present form remain difficult for people to grasp and thus respond effectively to an approaching flood. It is difficult to interpret the meaning of a water-level rise by 6 m for a person who is illiterate and resides far away from the river (Chowdhury 2005). This is one of the great deficiencies of the existing warning system. In Dhaka, flood warnings are available for major rivers such as Buriganga and Lakhya without any spatial detail. In such a situation, it is tricky to realize the connotation of the danger level (DL) for most of the megacity dwellers and prepare them for evacuation (Dewan 2006) (Plate 3.10). Therefore, the local context of flooding with plain language could be of significant assistance (Mallick et al. 2005). However, this is virtually absent in the megacity.

3.4.7 *Sociopolitical Structure*

The political system in Bangladesh is extremely segregated, and each citizen faces various challenges in order to have a "risk-free" life from *everyday hazards to large-scale disasters* (Bull-Kamanga et al. 2003). The economic and governance systems are also affected because of such political "biasness," causing serious environmental degradation along with the increase of massive marginal population. As noted by Sanderson (2012), three major reasons generate marginalization in urban Bangladesh: poor-quality governance in policy formulation and urban management; weak instruments of governance delivery, rule of law and corruption; and social and economic exclusion. Consequently, most of the environmental problems, including increasing flood vulnerability, are human-induced (Alam and Rabbani 2007). While some urban dwellers enjoy all types of entitlements, marginalized groups face incredible challenges in coping with this segregated political economy. For instance, affluent people and the political elites have access to resources and high-quality urban amenities, whereas many urban dwellers are constantly struggling to run their life with limited entitlements. Due to such political, social, and economic fragmentation, many existing policies are difficult to implement in the case of Dhaka, especially when the issue is related to wealthy people and the political elites, and they also exclude marginal groups from development planning. This is clearly visible in the encroachment of expensive natural resources such as wetlands and floodplains (Alam and Ahmad 2010; Begum 2007; Gain 2002). As a result, the detailed area plan (DAP) that was devised in 2005 to ameliorate environmental degradations, including flood vulnerability, could not be ratified until now, although a number of attempts were made by the concerned ministry.

Due to the great disparity in the sociopolitical structure, the *vulnerability gap* (Johnson 2010) is widening with the progress of time, putting more people and property at a greater risk of floods.

3.4.8 Governance Quality

Political stability is perhaps one of the most important prerequisites for good governance in a country (Kaufmann et al. 2010). Unfortunately, Bangladesh is a highly politically unstable country, which is seriously affecting all facets of life, including effective flood management. Although Bangladesh developed good emergency management systems that reduce disaster fatalities (Osti et al. 2011), corruption, nepotism, and a lack of transparency among responsible organizations are important obstacles in reducing flood risk in the Dhaka megacity (Barua and van Ast 2011). For example, many of the wetlands and low-lying areas are constantly being grabbed by real estate companies despite the existence of a number of legislations such as the Wetland Protection Act 2000. Good governance is also hampered by a lack of coordination among organizations involved in disaster management. The degree of conflict between the organizations involved is shown in the following paragraph:

> *'DCC claimed that the DWASA has broken the DCC drainage channels while maintaining the storm sewer lines. The channels have been filled up with mud. On the contrary, DWASA is claiming that DCC has unauthorized connection with storm sewer line. This creates the blockage in the storm sewer line and storm water cannot drain out through these. The DCC has approximately 2,500 km surface drains around the city, which are not being maintained for a long time. This is causing blockage in the storm sewer and therefore water logging in the city. The coordination meeting is taking place regularly but implementation is not happening (Datta 2008 in Barua and van Ast 2011).'*

As 14 ministries are directly or indirectly involved in Dhaka's flood management, a lack of coordination (Mohit 1992) among these organizations acts as an auxiliary factor to the rise of flood vulnerability, which was clearly evident in the previous flood events (Faisal et al. 2003; IFCDR 1998).

3.4.9 *Impact of Climate Change*

Many studies have demonstrated that Bangladesh would be the worst victim of climate change despite its tiny contribution to global greenhouse gas emissions (Senga 2004). The effects of climate change are projected to be the highest on water-related disasters, particularly floods (Mirza 2011). Climate change and widespread urban growth may intensify flooding in the coming years (Roy 2009). Although long-term annual rainfall patterns show a subtle variation, monsoonal rainfalls show an increasing tendency (Alam and Rabbani 2007), indicating that the flood problem could be aggravated under climate variability. Low elevation, rapid population growth—particularly slum dwellers—and limited adaptive capacity could put more people and property at risk of flood in the event of climate change (Balica et al. 2012; Senga 2004) if effective countermeasures are not taken.

3.4.10 *Absence of Risk Communication Tools*

Despite recurrent inundation during monsoons, hazard assessments, the mapping of social and physical vulnerability, and risk assessment in Bangladesh continue to be largely unexplored (Paul 1997). For instance, two major studies have been published (Nishat et al. 2000; Ali et al. 2002) regarding the repercussions of the 1998 floods, and one study has been published on the 2004 flood (Siddiqui and Hossain 2004); however, none of the papers listed in these books considered geospatial techniques for hazard and risk mapping. Moreover, historical hazard and risk maps are lacking, which hinders the ability to predict likely patterns of risk for this rapidly growing megacity (Huq 1999). Although BWDB (Bangladesh Water Development Board) regularly delineates flood areas based on a type of damage and subjective judgment, there are no precise numbers of informed flood areas (Jakobsen et al. 2005). Additionally, flood maps are medium-small scale, inaccessible to the public, and only used for administrative and forecasting purposes (Jha et al. 2012). Current flood response and recovery capabilities are mostly reactive and are inadequate in tackling catastrophic events. Thus, following the 1998 devastation, the water experts, urban planners, and key policy makers of the country have prioritized flood monitoring, hazard assessment, and disaster relief. They have underscored the need for the accurate mapping of flood-prone areas (Islam 2000), the assessment of hazards (Nishat 1998), and the development of flood risk maps (Hossain 1998). A flood hazard and risk map was prepared according to administrative unit by considering the 1998 flood with geospatial techniques (Dewan et al. 2006, 2007a, b); however, a particular event hazard map appears to be inadequate for the efficient management of recurrent flood disasters (Hoque et al. 2011). Moreover, a flood risk map without the social dimension analyzed may not provide an accurate picture of flood risk for a given area (Wu et al. 2002).

3.5 Summary

This chapter described the physical and socio-environmental characteristics of the Dhaka megacity. It revealed that Dhaka's population is growing rapidly, which is in turn accelerating the spatial expansion of the city. The growth of Dhaka is mostly unplanned and ill-structured, leading to the rise of impervious surfaces and slum populations, which has considerable implications for future floods.

Historical floods and associated damage was investigated, and it was found that flood losses are mounting. Although floods that occurred in the 1970s were not significantly different from those that occurred in the recent past, from a hydrological point of view, increasing flood damage in Dhaka can primarily be attributed to the growing occupancy of flood-prone areas and the development of urban infrastructures without considering the hydrological responses of land-use change. Other factors that influence Dhaka's vulnerability to floods include inadequate and dilapidated drainage infrastructures, high population density, extreme poverty, inequity of wealth distribution, lack of coordination among agencies, dearth of floodplain zoning, deplorable governance system, limited recovery and response capabilities, and poor performance of flood control works.

Although various adaptation strategies have been proposed (GoB 2005), their success in achieving efficient flood management depends on many issues, including restricting urban expansion on floodplains, particularly in Dhaka. As urban planning and related developments are based on traditional policies, many of these adaptation strategies are not able to work for two reasons. First, the ever-increasing population together with widespread poverty and inequality in resource distribution consistently forces thousands of people to settle in dangerous locations. Second, the city already has an *adaptation deficit* (Burton 2005/2006), as non-climatic factors such as unplanned and piecemeal urban expansions are consistently increasing exposure to flood hazards. If this trend continues, water-related disasters are likely to be exacerbated, suggesting greater uncertainty for the sustainability of Dhaka as a megacity.

References

Afsar R (2000) Rural-urban migration in Bangladesh: causes, consequences and challenges. University Press Limited, Dhaka

Ahmed SU (1986) Dhaka: a study in urban history and development 1840–1921. Curzon Press Limited, London

Ahmed SU, Gotoh K, Hossain S (2006) Integration of remote sensing and economic valuation techniques to find the causes of water logging problem in Dhaka City, Bangladesh. In: Proceedings of the 13 ARSPC conference, Australia, CD-ROM

Ahmed B, Ahmed R (2012) Modeling urban land cover growth dynamics using multi-temporal satellite images: a case study of Dhaka, Bangladesh. ISPRS Int J Geo-Inf 1(1):3–31

Ahmed KM, Nishigaki M, Dewan AM (2005) Constraints and issues on sustainable groundwater exploitation in Bangladesh. J Groundw Hydrol 47(2):163–179

Ahmed SJ, Bramley G, Dewan AM (2012) Exploratory growth analysis of a megacity through different spatial metrics: a case study on Dhaka, Bangladesh (1960–2005). URISA J 24(1):9–24

Akbar HMD, Minnery JR, Horen BV, Smith P (2007) Community water supply for the urban poor in developing countries, the case of Dhaka, Bangladesh. Habitat Int 31(1):24–35

Aktar Z, Fakhruddin ANM, Ahmed MSU, Nasreen T, Parveen S (2009) Assessment and treatment of the supplied pipeline water of selected areas of Dhaka City. Bangladesh J Sci Ind Res 44(3):335–340

Alam MJ, Ahmad MM (2010) Analysing the lacunae in planning and implementation: spatial development of Dhaka city and its impact upon the built environment. Int J Urban Sustain Dev 2(1–2):85–106

Alam M, Rabbani MG (2007) Vulnerabilities and responses to climate change for Dhaka. Environ Urban 19(1):81–97

Alam BN, Damole LN, Wickramanyake E (1996) Effects of flood mitigation measure: lessons from Dhaka flood protection project, Bangladesh. Asian Profile 24(6):511–524

Ali MA, Seraj SM, Ahmed S (eds) (2002) Engineering concerns of flood. Bangladesh University of Engineering and Technology, Dhaka

Angeles G, Lance P, Barden-O'Fallon J, Islam N, Mahbub AQM, Nazem NI (2009) The 2005 census and mapping slums in Bangladesh: design, select results and application. Int J Health Geogr 8:32

Bala SK, Islam AKMS, Chowdhury JU, Rahman MR, Haque MA, Khan MSA, Salehin M (2009) Performance of flood control works around Dhaka city during major floods in Bangladesh. A paper presented in the 2nd international conference on water and flood management, Dhaka

Balica SF, Wright NG, van der Muelen F (2012) A flood vulnerability index for coastal cities and its use in assessing climate change impacts. Nat Hazard. doi:10.1007/s11069-012-0234-1

Bangladesh Bureau of Statistics (BBS) (1991) Population census – 1991. Community series. Bangladesh Bureau of Statistics, Ministry of Planning, Dhaka

Bangladesh Bureau of Statistics (BBS) (2001) Population census – 2001. Zila series. Bangladesh Bureau of Statistics, Ministry of Planning, Dhaka

Bangladesh Bureau of Statistics (BBS) (2003) Population census – 2001. Community series. Bangladesh Bureau of Statistics, Ministry of Planning, Dhaka

Bangladesh Bureau of Statistics (BBS) (2005) Statistical yearbook of Bangladesh. Bangladesh Bureau of Statistics, Ministry of Planning, Dhaka

Bangladesh Bureau of Statistics (BBS) (2008) Population census-2001. National series, Vol 3, Urban Area report. Bangladesh Bureau of Statistics, Ministry of Planning, Dhaka

Bangladesh Bureau of Statistics (BBS) (2010) Compendium of environment statistics of Bangladesh 2009. Bangladesh Bureau of Statistics, Ministry of Planning, Dhaka

Bangladesh Bureau of Statistics (BBS) (2011) Population census – 2011. Preliminary report. Bangladesh Bureau of Statistics, Ministry of Planning, Dhaka

Bangladesh Bureau of Statistics (BBS) (2012) Population and housing census – 2011. Community report, Dhaka, Narayanganj and Gazipur Zila. Bangladesh Bureau of Statistics, Ministry of Planning, Dhaka

Banglapedia (2006) http://www.banglapedia.org/httpdocs/HT/D_0151.HTM. Accessed 5 June, 2011

Barua S, van Ast JA (2011) Towards interactive flood management in Dhaka, Bangladesh. Water Policy 13(5):693–716

Begum A (2007) Urban housing as an issue of redistribution through planning? The case of Dhaka city. Soc Policy Adm 41(4):410–418

Benson C, Clay EJ (2003) Disasters, vulnerability and the global economy. In: Kreimer A, Arnold M, Carlin A (eds) Building safer cities: the future of disaster risk. The World Bank, Washington, DC, pp 3–32
Bhattarai K, Conway D (2010) Urban vulnerabilities in the Kathmandu valley, Nepal: visualizations of human/hazard interactions. J Geogr Inf Syst 2(2):63–84
Braun B, Aßheuer T (2011) Floods in megacity environments: vulnerability and coping strategies of slum dwellers in Dhaka/Bangladesh. Nat Hazard 58(2):771–787
Bromley R, Hall P, Dutt A, Mookherjee D, Benhart JE (1989) Regional development and planning. In: Gaile GL, Willmott CJ (eds) Geography in America. Merrill Publishing Company, Columbus
Bull-Kamanga L, Diagne K, Lavell A, Leon E, Lerise F, MacGregor H, Maskrey A, Meshack M, Pelling M, Reid H, Satterthqaite D, Songsore J, Westgate K, Yitambe A (2003) From everyday hazards to disasters: the accumulation of risk in urban areas. Environ Urban 15(1):193–204
Burton I (2005/2006) Adapt and thrive: options for reducing the climate change adaptation deficit. Policy options, December 2005 – January 2006, pp 33–38
Center for Urban Studies (CUS) (1990) The urban poor in Bangladesh. CUS, Dhaka
Center for Urban Studies (CUS), National Institute of Population, Research and Training (NIPORT), Measure Evaluation (2006) Slums in urban Bangladesh: mapping and census 2005. CUS, Dhaka
Choudhury NY (2001) Approaches in the flood hazard studies in Bangladesh: a review. J Inst Bangladesh Stud 14(1):63–68
Chowdhury MR (2000) As assessment of flood forecasting in Bangladesh: the experience of the 1998 flood. Nat Hazard 22(2):139–163
Chowdhury MR (2003) The impact of 'Greater Dhaka Flood Protection Project' (GDFPP) on local living environment–the attitude of the floodplain residents. Nat Hazard 29(3):309–324
Chowdhury MR (2005) Consensus seasonal flood forecasts and warning response system (FFWRS): an alternate for non-structural flood management in Bangladesh. Environ Manage 35(6):716–725
Chowdhury FJ, Amin ATM (2006) Environmental assessment in slum improvement programs: some evidence from a study on infrastructure projects in two Dhaka slums. Environ Impact Assess Rev 26(6):530–552
Chowdhury AM, Faruqui S (1991) Physical growth of Dhaka City. In: Ahmed SU (ed) Dhaka: past, present and future. The Asiatic Society of Bangladesh, Dhaka, pp 43–61
Chowdhury MR, Sato T (2000) The impact of household characteristics on flood damage – a case study. J Floodplain Manage 1(2):35–56
Chowdhury JU, Rahman R, Bala SK, Islam AKMS (1998) Impact of 1998 flood on Dhaka City and performance of flood control works. Institute of Flood Control and Drainage Research, Bangladesh University of Engineering and Technology, Dhaka
Cutter SL, Mitchell JT, Scott MS (2000) Revealing the vulnerability of people and places: a case study of Georgetown County, South Carolina. Ann Assoc Am Geogr 90(4):713–737
Dewan AM (2006) Flood risk in Greater Dhaka: an empirical analysis. Geogr Environ J 5(1):160–171 (in Bengali)
Dewan AM, Yeboah KK, Nishigaki M (2006) Flood hazard delineation in Greater Dhaka, Bangladesh using integrated GIS and remote sensing approach. Geocarto Int 21(2):33–38
Dewan AM, Corner RJ (2012) The impact of land use and land cover changes on land surface temperature in a rapidly urbanizing megacity, IGARSS 2012, Munich, Germany
Dewan AM, Yamaguchi Y (2007). GIS and remote sensing for mapping and monitoring the effect of land use/cover changes on flooding in Greater Dhaka of Bangladesh. Paper presented at RSPSOC-2007, Newcastle University, Newcastle upon Tyen, UK
Dewan AM, Yamaguchi Y (2008) Effects of land cover changes on flooding: example from Greater Dhaka of Bangladesh. Int J Geoinf 4(1):11–20
Dewan AM, Yamaguchi Y (2009a) Land use and land cover change in Greater Dhaka, Bangladesh: using remote sensing to promote sustainable urbanization. Appl Geogr 29(3):390–401
Dewan AM, Yamaguchi Y (2009b) Using remote sensing and GIS to detect and monitor land use and land cover change in Dhaka Metropolitan of Bangladesh during 1960–2005. Environ Monit Assess 150(1–4):237–249

Dewan AM, Islam MM, Kumamoto T, Nishigaki M (2007a) Evaluating flood hazard for land-use planning in Greater Dhaka of Bangladesh using remote sensing and GIS techniques. Water Resour Manage 21(9):2101–2116

Dewan AM, Kabir MH, Islam MM, Kumamoto T, Nishigaki M (2007b) Delineating flood risk areas in Greater Dhaka of Bangladesh using geoinformatics. Georisk: Assess Manage Risk Eng Syst Geohazard 1(4):190–201

Dewan AM, Kabir MH, Nahar K, Rahman MZ (2012a) Urbanization and environmental degradation in Dhaka metropolitan area of Bangladesh. Int J Environ Sustain Dev 11(2):118–146

Dewan AM, Yamaguchi Y, Rahman MZ (2012b) Dynamics of land use/cover changes and the analysis of landscape fragmentation in Dhaka Metropolitan, Bangladesh. GeoJournal 77(3):315–330

Disaster Management Bureau (DMB) (1998) Report on Bangladesh flood 1998: chronology, damages and responses. DMB, Dhaka

Dutt AK, Noble AG (2004) Urban development in South Asia. In: Dutt AK, Noble AG, Venugopal G, Subbiah H (eds) Challenges to Asian urbanization in the 21st century. Kluwer Academic Publishers, Dordrecht, pp 256–273

Faisal IM, Kabir MR, Nishat A (1999) Non-structural flood mitigation measures for Dhaka City. Urban Water 1(2):145–153

Faisal IM, Kabir MR, Nishat A (2003) The disastrous floods of 1998 and long term mitigation strategies for Dhaka City. Nat Hazard 28(1):85–99

Flood Action Plan (FAP) 8A (1991) Master plan study for Greater Dhaka protection project. Japan International Cooperation Agency, Dhaka

Gain P (2002) Bangladesh environment facing the 21st century, 2nd edn. Society for Environment and Human Development (SEHD), Dhaka

Godschalk DR (2003) Urban hazard mitigation: creating resilient cities. Nat Hazard Rev 4(3):136–143

Government of Bangladesh (GoB) (2005) National adaptation programme of action. Ministry of Environment and Forest, Government of Bangladesh, Dhaka

Griffiths P, Hostert P, Gruebner O, van der Linden S (2010) Mapping megacity growth with multi-sensor data. Remote Sens Environ 114(12):426–439

Haq KA (2006) Water management in Dhaka. Int J Water Resour Dev 22(2):291–311

Haque AN, Grafakos S, Huijsman M (2012) Participatory integrated assessment of flood protection measures for climate adaptation in Dhaka. Environ Urban 24(1):197–213

Harris AM, Chowdhury F, Begum YA, Khan AI, Faruque ASG, Svennerholm A-M, Harris JB, Ryan ET, Cravioto A, Calderwood SB, Qadri F (2008) Shifting prevalence of major diarrheal pathogens in patients seeking hospital care during floods in 1998, 2004, and 2007 in Dhaka, Bangladesh. Am J Trop Med Hyg 79(5):708–714

Hasan S, Mulamoottil G (1994) Environmental problems of Dhaka City. Cities 11(3):195–200

Hasnat MA (2004) Flood disaster management in Dhaka city. In: Siddiqui KU, Hossain ANHA (eds) Options for flood risk and damage reduction in Bangladesh. The University Press Limited, Dhaka, pp 121–129

Hewitt K (1997) Regions of risk: a geographical introduction to disasters. Longman, Essex

Hoque R, Nakayama D, Matsuyama H, Matsumoto J (2011) Flood monitoring, mapping and assessing capabilities using RADARSAT remote sensing. GIS Ground Data Bangladesh 57(2):525–548

Hossain MM (1998) Bangladesh floods and its management initiatives. Steps Towards Dev 3:12–19

Hossain S (2006) Social characteristics of a megacity: a case of Dhaka city: Bangladesh. TASA conference, Perth, Australia, 4–7 December

Hossain AMMM, Rahman S (2011) Hydrography of Dhaka city catchment and impact of urbanization on water flows: a review. Asian J Water Environ Pollut 8(4):27–36

Hossain AMMM, Rahman S (2012) Impact of land use and urbanization activities on water quality of the megacity, Dhaka. Asian J Water Environ Pollut 9(2):1–9

Hunter WW (1877) A statistical account of Bengal, vol V. Trubner and Co, London

Huq S (1999) Environmental hazards in Dhaka. In: Mitchell JK (ed) Crucibles of hazards: megacities and disasters in transition. United Nations University, Tokyo, pp 119–137
Huq S, Alam M (2003) Flood vulnerability of Dhaka City and its management. In: Kreimer A, Arnold M, Carlin A (eds) Building safer cities: the futures of disaster risk. The World Bank, Washington, DC, pp 121–145
Hye A (2000) Effects on livelihood. In: Nishat A, Reazuddin R, Amin R, Khan AR (eds) The 1998 flood: impact on environment of Dhaka City. Department of Environment, Dhaka, pp 201–216
Institute of Epidemiology, Disease Control and Research (IEDCR) (2004) Flood 2004 and diarrhoea. Report on outbreak investigation. IEDCR, Dhaka
Institute of Flood Control and Drainage Research (IFCDR) (1998) Impact of 1998 flood on Dhaka City and performance of flood control works. IFCDR, Dhaka
Islam N (1991) Dhaka in 2025 AD. In: Ahmed S (ed) Dhaka: past, present and future. The Asiatic Society of Bangladesh, Dhaka, pp 570–583
Islam N (1996) Dhaka from city to megacity: perspectives on people, places, planning and development issues. Urban Studies Program, Department of Geography, University of Dhaka, Dhaka
Islam N (1998) Flood'98 and the future of urban settlements in Bangladesh. In: National seminar on flood'98 and management of floods in future, jointly organized by National Committee for Relief and Rehabilitation and Dhaka University, Dhaka, December 8 1998
Islam N (1999) Urbanization, migration and development in BD: recent trends and emerging issues. CPD-UNFPA series 1. Centre for Policy Dialogue, Dhaka, p 14
Islam N (2000) Housing damage in Dhaka City due to the 1998 flood. In: Nishat A, Reazuddin R, Amin R, Khan AR (eds) The 1998 flood: impact on environment of Dhaka City. Department of Environment, Dhaka, pp 95–103
Islam N (2005a) Dhaka now: contemporary urban development. Bangladesh Geographical Society, Dhaka
Islam N (2005b) Natural hazards in Bangladesh: studies in perception, impact and coping strategies. Disaster Research, Training and Management Center, University of Dhaka, Dhaka
Islam N (2005c) Improving the lives of slum dwellers in Bangladesh. CUS Bulletin on Urbanization and Development, no. 48. Center for Urban Studies, Dhaka
Islam I (2009) Wetlands of Dhaka metro area: a study from social, economic and institutional perspectives. A H Development Publishing House, Dhaka
Islam MS, Ahmed R (2011) Land use change prediction in Dhaka city using GIS aided Markov chain modeling. J Life Earth Sci 6:81–89
Islam N, Mahbub AQM (1988) Slums and squatters in Dhaka City 1988. Centre for Urban Studies (CUS), Dhaka
Islam MM, Chou FN-F, Kabir MR (2011) Feasibility and acceptability study of rainwater use to the acute water shortage areas in Dhaka city, Bangladesh. Nat Hazard 56(1):93–111
Jabeen H, Johnson C, Allen A (2010) Built-in resilience: learning from grassroots coping strategies for climate variability. Environ Urban 22(2):415–431
Jahan S (2000) Coping with flood: the experience of the people of Dhaka during the 1998 flood disaster. Aust J Emerg Manage Spring:16–20
Jakobsen F, Hoque AKMZ, Paudyal GN, Bhuiyan MS (2005) Evaluation of the short-term forcing the monsoon river floods in Bangladesh. Water Int 30(3):389–399
Jha AK, Bloch R, Lamond J (2012) Cities and flooding: a guide to integrated urban flood risk management for the 21st century. The World Bank, Washington, DC
Johnson C (2010) Urban disaster trends. In: World disasters report – focus on urban risk. International Federation of Red Cross and Red Crescent Societies, Geneva, pp 31–51
Kafiluddin AKM (2000) Impact study on health. National Institute of Preventive and Social Medicine (NIPSOM), Dhaka
Kamal ASMM, Midorikawa S (2004) GIS-based geomorphological mapping using remote sensing data and supplementary geoinformation a case study of the Dhaka City area, Bangladesh. Int J Appl Earth Observ Geoinf 6(2):111–125
Kamal MM, Malmgren-Hansen A, Badruzzaman ABM (1999) Assessment of pollution of the river Buriganga, Bangladesh using a water quality model. Water Sci Technol 40(2):129–136

Karim CS, Hye JMA, Ahmed N, Shahadat F (2005) Study of the development of water body at Ashulia and its environmental impact. Nuclear Sci Appl 14(2):21–33

Karn SK, Harada H (2001) Surface water pollution in three urban territories of Nepal, India and Bangladesh. Environ Manage 28(4):483–496

Kaufmann D, Karry A, Mastruzzi M (2010) The worldwide governance indicators: methodology and analytical issues. Policy Research working paper 5430. The World Bank Institute, Washington, DC

Khan NI (2000) Temporal mapping and spatial analysis of land transformation due to urbanization and its impact on surface water system: a case from Dhaka Metropolitan area. In Arch Photogramm Remote Sens. Available at http://www.isprs.org/proceedings/XXXIII/congress/part7/598_XXXIII-part7.pdf

Khan NI (2001) Assessment of water logging conditions using integrated GIS and remote sensing techniques: a study on Dhaka Mega City. Orient Geogr 45(2):41–54

Khan MSA (2006) Stormwater flooding in Dhaka city: causes and consequences. J Hydrol Meteorol 3(1):77–85

Khan MSA (2008) Disaster preparedness for sustainable development in Bangladesh. Disast Prev Manage 17(5):662–671

Lall SV, Deichmann U (2012) Density and disasters: economics of urban hazard risk. World Bank Res Observ 27(1):74–105

Maathuis BHP, Mannaerts CMM, Khan NI (1999) Evaluating urban storm water drainage using GIS and RS techniques – a case study of Dhaka City, Bangladesh. Geocarto Int 14(4):21–32

Mallick DL, Rahman A, Alam A, Juel ASM, Ahmad AN, Alam SS (2005) Bangladesh floods: a shift from disaster management towards disaster preparedness. IDS Bull 36(4):53–71

Maroof FBA, Hadi DA, Khan AH, Chowdhury AH (1986) Cadmium and Zinc concentration in drinking water supplies of Dhaka City, Bangladesh. Sci Total Environ 53(3):233–238

Masood M, Takeuchi K (2012) Assessment of flood hazard, vulnerability and risk of mid-eastern Dhaka using DEM and 1D hydrodynamic model. Nat Hazard. doi:10.1007/s11069-011-0060-x

Meyer V, Scheuer S, Haase D (2009) A multicriteria approach for flood risk mapping exemplified at the Mulde river, Germany. Nat Hazard 48(1):17–39

Miah MM, Bazlee BL (1968) Some aspects of geomorphology of the Madhupur Tract. Orient Geogr 12(1):39–48

Miah MM, Mohit MA (1996) Factors affecting the performance of flood control, drainage and irrigation projects in Bangladesh: a case study, J Civil Eng, IEB, Bangladesh CE24(2):147–154

Mirza MMQ (2011) Climate change, flooding in South Asia and implications. Reg Environ Chang 1(supp):95–107

Mitchell JK (1999a) Natural disasters in the context of mega-cities. In: Mitchell JK (ed) Crucibles of hazard: mega-cities and disasters in transition. United Nations University, Tokyo, pp 15–55

Mitchell JK (1999b) Megacities and natural disasters: a comparative analysis. GeoJournal 49(2):137–142

Mohit MA (1992) Institutional arrangement for the development of Dhaka Metropolitan Area: problems and issues. In: Islam N, Chowdhury AI (eds) Urban land management in Bangladesh. Ministry of Land, Dhaka, pp 99–108

Mohit MA, Akhter S (2002) Delineation of flood damaged zones of Dhaka City based on the 1998 flood by using GIS. In: Ali MA, Seraj SM, Ahmed S (eds) Engineering concerns of flood. Bangladesh University of Engineering and Technology, Dhaka, pp 303–318

Nishat A (1998) A discussion on flood management in Bangladesh. In: The 1998 Deluge-developing coping capacities, special workshop on the 1998 flood. Nagorik Durjog-Mokabela Uddogh, Dhaka

Nishat A, Reazuddin R, Amin R, Khan AR (eds) (2000) The 1998 Flood: impact of environment of Dhaka City. Department of Environment, Dhaka

Okubo K, Khan MSA, Hassan Q (2010) Hydrological processes of adsorption, sedimentation and infiltration into the lake bed during the 2004 urban flood in Dhaka city, Bangladesh. Environ Earth Sci 60(1):95–106

Osti R, Hishinuma S, Miyake K, Inomata H (2011) Lessons learned from statistical comparison of flood impact factors among southern and eastern Asian countries. J Flood Risk Manage 4(3):203–215

Paul BK (1997) Flood research in Bangladesh in retrospect and prospect: a review. Geoforum 28:121–131

Pelling M (2003) The vulnerability of cities. Earthscan, London

Rahman S, Hossain F (2008) Spatial assessment of water quality in peripheral rivers of Dhaka city for optimal location of water intake point. Water Resour Manage 22(3):377–391

Rahman M, Karim A (1995) Impact of Dhaka city flood protection embankment on surface water quality with respect to tannery waste disposal. Int J Pollut Res 14(2):145–156

Rahman MR, Azad AK, Nahar SB, Ali A, Hossain M, Jahan Y, Ara R, Hoque MM (2004) The impact of floods on health and hygiene in rural and urban areas and mitigation options. In: Siddiqui KU, Hossain ANH (eds) Options for flood risk and damage reduction in Bangladesh. The University Press Limited, Dhaka, pp 383–398

Rana MMP (2010) Urbanization and sustainability: challenges and strategies for sustainable urban development in Bangladesh. Environ Dev Sustain 13(1):237–256

Rashed T, Weeks J (2003) Assessing vulnerability to earthquake hazards through spatial multi-criteia analysis of urban areas. Int J Geogr Inf Sci 17(6):547–576

Rashed R, Weeks J, Couclelis H, Herold M (2007) An integrative GIS and remote sensing model for place-based urban vulnerability analysis. In: Mesev V (ed) Integration of GIS and remote sensing. Wiley, Chichester, pp 199–224

Rashid SF (2000) The urban poor in Dhaka city: their struggles and coping strategies during the floods of 1998. Disasters 24(3):240–253

Rashid H, Hunt LM, Haider W (2007) Urban flood problems in Dhaka, Bangladesh: slum residents' choices for relocation to flood-free areas. Environ Manage 40(1):95–104

Rasid H, Mallik A (1996) Living on the edge of stagnant water: an assessment of environmental impacts of construction-phase drainage congestion along Dhaka City flood control embankment, Bangladesh. Environ Manage 20(1):89–98

Rizvi SNH (1969) East Pakistan district gazetteers-Dacca. East Pakistan Government Press, Dacca

Roy M (2009) Planning for sustainable urbanization in fast growing cities: mitigation and adaptation issues addresses in Dhaka, Bangladesh. Habitat Int 33(3):276–286

Sanderson D (2012) Building livelihoods to reduce risk among the most marginalized in urban areas: strategic approaches from Dhaka. Environ Hazard 11(2):112–122

Senga R (2004) Natural and unnatural disasters: the relative vulnerabilities of Southeast Asian megacities to climate change. World Wide Fund for Nature (WWF), Switzerland. www2.lse.ac.uk/IDEAS/publications/reports/pdf/SR004/WWF.pdf. Accessed 15 Mar, 2012

Siddique QI, Chowdhury MH (1998) Flood-98: losses and damages. In: National seminar on flood'98 and management of floods in future, jointly organized by National Committee for Relief and Rehabilitation and Dhaka University, Dhaka, 8 December, 1998

Siddiqui KU, Hossain ANHA (2004) Options for flood risk and damage reduction in Bangladesh. The University Press Limited, Dhaka

Siddiqui K, Qadir SR, Alamgir S, Haq S (1993) Social formation in Dhaka City. University Press Limited, Dhaka

Siddiqui K, Ghosh A, Bhowmik SK, Siddiqi SA, Mitra M, Kapuria S, Rajan N, Ahmed J (2004) Megacity governance in South Asia. University Press Limited, Dhaka

Sohel KMA, Chowdhury MAI, Ahmed MF (2003) Surface water quality in and around Dhaka City. J Water Supply: Res Technol 52(2):141–153

Stalenberg B, Vrijling H (2009) The battle of Tokyo and Dhaka against floods. Built Environ 35(4):471–491

Sultana MS, Islam GMT, Islam Z (2009) Pre- and post-urban wetland area in Dhaka City, Bangladesh: a remote sensing and GIS analysis. J Water Resour Prot 1:414–421

Taubenböck H, Post J, Roth A, Zosseder K, Strunz G, Dech S (2008) A conceptual vulnerability and risk framework as outline to identify capabilities of remote sensing. Nat Hazard Earth Syst Sci 8(3):409–420

Tawhid KG (2004) Causes and effects of water logging in Dhaka City, Bangladesh. TRITA-LWR master thesis, Department of Land and Water Resource Engineering, Royal Institute of Technology, Sweden

UN-Habitat (2003) The challenges of slums: global report on human settlements. Earthscan, London

United Nations (2012) World population prospects: the 2010 revision, vol 1. United Nations Department of Economic and Social Affairs, New York. Available at: http://eas.un.org/undp/wpp/Documentation/publications.html. Accessed 1 Apr, 2012

Varis O, Biswas AK, Tortajada C, Lundqvist J (2006) Megacities and water management. Int J Water Resour Dev 22(2):377–394

World Bank (2007) Dhaka: improving living conditions for the urban poor. Sustainable Development Unit, South Asia Region, Report No. 35824-BD

Wu SY, Yarnal B, Fisher A (2002) Vulnerability of coastal communities to sea-level rise: a case study of Cape May County, New Jersey, USA. Clim Res 22:255–270

Zahid A, Balke KD, Hassan MQ, Flegr M (2006) Evaluation of aquifer environment under Hazaribagh leather processing zone of Dhaka city. Environ Geol 50(4):495–504

Chapter 4
Spatial and Temporal Distribution of Floods

Abstract Flood mapping and monitoring was performed using both Landsat and Radarsat data from 1988 to 2009. Multi-temporal satellite data were digitally classified using a threshold algorithm to determine the spatial and temporal distribution of floods. In addition, flood depths were calculated using the highest water-level data with a DEM. Synthetic depth–damage curves were prepared for three housing categories and for major roads. Flood maps were evaluated using classified satellite images with ground truth data. It was found that 25% of the study area is flooded every year which could rise to more than 40% during abnormal events such as those that occurred in 1988 and 1998. Flood damage estimations revealed that the greatest damage occurred to *katcha* houses rather than *semi-pucca* and *pucca* houses. Variable accuracies were found for satellite-derived flood maps; however, the overall accuracy was highest for radar-based classifications.

4.1 Introduction

Accurate delineation of flood-prone areas using remote sensing (RS) has received considerable attention in the past few decades due to the drawbacks associated with the traditional survey method, including the erroneous demarcation of flood-prone areas (Smith 1997). Data from earth-observing satellites is of particular importance for providing current and reliable surface information, and the data are deemed suitable for developing countries, as lack of timely information is an impediment for assessing risks from naturally occurring phenomena such as floods.

To formulate effective flood management strategies, the most crucial step is to map the areas that are most liable to flood. These data can then be fed into risk assessment models (Gillespie et al. 2007; McMillan et al. 2006) to identify social inequality patterns relating to flood risk exposure (Walker and Burningham 2011). Flood damage assessments provide another important aspect of the flood alleviation scheme (Wind et al. 1999). A flood map is of little help unless it is combined with

A.M. Dewan, *Floods in a Megacity: Geospatial Techniques in Assessing Hazards, Risk and Vulnerability*, Springer Geography, DOI 10.1007/978-94-007-5875-9_4,

land-use information for damage assessment (Consuegra et al. 1995). It is believed that 40% of the damage associated with floods can be reduced by integrating flood maps with land-use information in a given area (Marco 1992). In addition, precise flood maps combined with damage information can be used to develop comprehensive relief efforts (Corbley 1993), generate spatially accurate hazard maps (Horritt et al. 2001), and to determine the deficiencies of existing flood control works that support the development of flood countermeasures (Boyle et al. 1998).

As discussed in Chap. 1, there are four types of urban flooding. Although both pluvial and fluvial floods are present in Dhaka, this study only focuses on the latter type because it was not feasible to map the spatial extent of localized flooding using moderate-resolution RS data. This chapter primarily deals with the mapping of flood-prone areas in the study site (see Fig. 3.1). Section 4.2 describes the methods and materials used to estimate flooding from 1988 to 2009. The calculation of flood depth in the same period is also illustrated. The accuracy of flood maps is explained in Sect. 4.3, and the spatial and temporal distribution of floods is presented in Sect. 4.4. The result of the accuracy of flood maps is elucidated in Sect. 4.5. An attempt was made to develop synthetic depth–damage curves by integrating derived depth maps with housing and road network information within a GIS (Sect. 4.6). A discussion section is presented in Sect. 4.7.

4.2 Analytical Techniques

4.2.1 Data

As discussed in Chap. 3, Dhaka experienced four major floods in 1988, 1998, 2004, and 2007. Images for these floods, along with other flood events, have been acquired (see Table 4.1) that included both optical and microwave data. Two Landsat TM scenes and eight Radarsat SAR images were used to estimate floods. All images being considered represent the peak floods of the respective year, except for 1988, and the acquisition of the images was decided on the basis of the water-level information of five major rivers and the availability of data in the archives of the Geo-Informatics and Space Technology Development Agency (GISTDA) of Thailand, the Space Research and Remote Sensing Organization (SPARRSO), and the Center for Environmental and Geographic Information Services (CEGIS) in Bangladesh. Due to the lack of cloud-free peak time images in 1988, the October 1988 image still represented a time closer to the peak flood (Chowdhury et al. 1998). As shown in Table 4.1, the radar data obtained for this study include both ascending and descending orbits at a single wavelength (C band) with fixed polarization (HH: horizontal–horizontal). Additionally, only the ScanSAR narrow beam mode was used because of its moderate spatial resolution and the unavailability of other beam modes in their archives. Note that the ScanSAR wide mode was available for a few flood years, but they are coarse in resolution (100 m). The incidence angle ranged between 31 and 46°. It should be noted that both extreme and normal floods were considered in this

Table 4.1 Attributes of satellite data

Image date	Orbit/ path /row	Spatial resolution (m)	Band	Polarization	Radarsat beam mode	Source
16 Oct 1988	137/43–44	28.5	1–7	–	–	GISTDA
26 Aug 1998	D[a]	50	C	H–H	ScanSAR narrow	SPARRSO
13 Aug 2000	A[b]	50	C	H–H	ScanSAR narrow	CEGIS
8 Aug 2001	A	50	C	H–H	ScanSAR narrow	CEGIS
22 Aug 2003	A	50	C	H–H	ScanSAR narrow	CEGIS
23 July 2004	A	50	C	H–H	ScanSAR narrow	SPARRSO
11 Aug 2005	A	50	C	H–H	ScanSAR narrow	CEGIS
3 Aug 2007	A	50	C	H–H	ScanSAR narrow	SPARRSO
10 July 2009	D	50	C	H–H	ScanSAR narrow	CEGIS

A[b] ascending, D[a] descending

study so a comparison could be made to detect the spatial distribution of floods in different years. Further, it would allow the recognition of the exact flood-prone areas during normal events, which may be of help for emergency planning of urban areas.

In addition to satellite data, ancillary information was acquired, including a land-use/land-cover map of the study area from February 2000 (Dewan and Corner 2012) to support the estimation of net inundated areas. Historical maps for major flood events were obtained from the Flood Forecasting and Warning Centre (FFWC) of the BWDB (Bangladesh Water Development Board) in the form of JPEGs, and they were subsequently integrated into GIS for accuracy assessment. The highest water-level (HWL) data of five rivers in Dhaka between 1973 and 2009 were obtained from BWDB. A DEM of the study area was obtained from the Survey of Bangladesh (SOB) to assist in the efficient analysis of flood hydrology. The DEM was interpolated to 30 m and used in conjunction with HWL data to derive flood depth maps in different years.

To support a community-based housing damage evaluation, a spatial database of the census tracts was created from the small area atlas of the Bangladesh Bureau of Statistics (BBS), the digital databases of CEGIS and SPARRSO. It should be noted that neither of the databases contained up-to-date digital census tract boundaries and around 2% of the census tracts were missing. That is, 25 new census districts were created between decennial census (1991–2001). To overcome this problem, the names of the 1991 census tracts were first matched with the 2001 names using the community series of BBS. Note that several census boundaries are divided from the existing boundaries due to the population increase. To create the polygons for the 2001 census, the digitized polygon shape files were re-referenced to the street grid, which is an up-to-date street network from the Detailed Area Plan (DAP) of the Capital Development Authority (RAJUK). Census units created between 1991 and 2001 were then demarcated and mapped out in the fields. A hard copy map was obtained from BBS which highlighted the road networks that were used to split the original census tracts for the 2001 population census. Further, a high-resolution mobile-mapping GPS device (Trimble Nomad 800GXE) was used to confirm the road locations. The editing of the census tracts was performed in ArcGIS (v. 10). The final census tract file consists of 1,212 polygon features, which include 441 rural communities known as *mauza* and 771 urban communities called *mahalla*. For simplicity, *mauza* and *mahalla* are expressed as "community" throughout the text, since they are known as community in the census geography of Bangladesh (BBS 2003). Other spatial data include housing types and road networks, which were obtained from the DAP database. It should be noted that road network data, from the DAP, were generated in 2004–2005 and thus were subject to an update. The road network data were updated using a high-resolution GeoEye image of 2010 to support infrastructural damage analysis for the 2007 flood. The location of embankments was obtained from CEGIS.

4.2.2 Satellite Data Preprocessing

The preprocessing of RS data includes georeferencing, mosaicking, and the pre-classification filtering of radar data.

4.2.2.1 Speckle Suppression of Radar Data

Speckle, which is an inherent problem to SAR images, appears as a granular texture and causes a noisy appearance accompanied by high standard deviations (Durand et al. 1987). To effectively use SAR data, speckle needs to be removed prior to classification; otherwise, its presence could hinder efficient classification (Nuesch 1982). Speckle reduction from SAR images requires a specialized filtering process to normalize the spikes. However, while speckle cannot be removed entirely from SAR, its effects can significantly be reduced by using filtering algorithms.

In this stage of image preprocessing, speckle filtering was applied before the georeferencing of all SAR images so the original image statistics would not be altered. There are few well-developed filtering methods to suppress the speckle (Henderson and Lewis 1998); however, they are not all equally applicable in the context of Bangladesh (FAP 19 1995). It was found that the gamma–MAP (maximum *a posteriori*) (Oliver 1991) filter performs better than other algorithms (Martin et al. 1996). Hence, the gamma-MAP filter was used to reduce the effects of speckle with various window sizes. It was realized that the application of a 5×5 window size was suitable for the area being investigated, keeping in mind that edge degradation can be a crucial factor for flood detection in urban areas (Jensen 1979). The result of filtering the speckle of SAR images is illustrated in Fig. 4.1.

4.2.2.2 Geocoding of Satellite Data

Geocoding involves the transformation of SAR data from ground/slant range geometry into map geometry so that comparisons with other datasets can be made independently. To integrate the datasets used in this study, the geocoding of SAR images was first

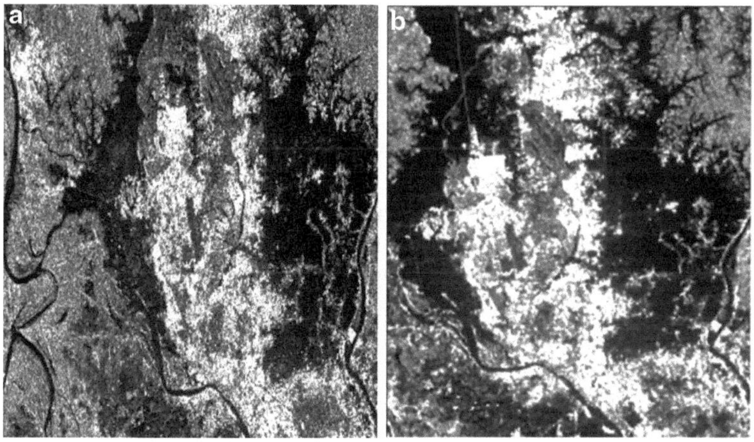

Fig. 4.1 Radarsat ScanSAR raw and filtered images (not to scale) (**a**) Raw image, (**b**) Gamma–MAP filtered image

performed using a geometrically corrected topographic map of 1997 as the reference. More than 130 ground control points (GCP) that were uniformly distributed over the area of interest were used to ensure the proper geometric correction of SAR data. A second-order polynomial fit was applied, and pixel values were resampled to 30 m using the cubic convolution technique. When the root mean square error (RMSE) was in the satisfactory range (<0.5 pixels), geometric correction was performed. The resulting RMSE ranged between 0.45 and 0.48 pixels. Similarly, Landsat TM data was geocoded using the same topographic map; however, an exception was that the first-order transformation with the nearest neighbor technique was used. A local projection system called Bangladesh Transverse Mercator (BTM) was used to georeference all satellite data. Upon georeferencing, they were mosaicked and then clipped using an area of interest (AOI) file derived from the study area's polygon file.

4.2.2.3 Mapping Flood Extent and Depth

The identification of water versus non-water from Landsat data encompasses a number of techniques (see Chap. 2). Landsat TM4, which yields very low reflectance in the reflected infrared region (0.76–0.90 μm), was found to be suitable in an early study (Bhavasar 1984); however, a few studies indicated that confusion could occur between water and asphalt areas in the TM4, which may increase the misclassification of flooded pixels (Baumann 1999). To overcome this problem, a mid-infrared band such as TM7 (2.08–2.35 μm) was incorporated with TM4, which is said to be a good alternative in which water versus non-water areas can easily be separated (Wang et al. 2002). They further noticed that TM5 (1.55–1.75 μm) could be utilized to discriminate floods from other land-use categories; however, TM7 provides superior results when used in combination with TM4.

In this study, a completely cloud-free TM image was used to estimate flooded areas during 1988. TM bands 4+7 were used and a threshold technique was adopted. Empirical threshold values were obtained from image spectral property, standard False Color Composite (FCC), and a flood depth map. Upon determining the cutoff value for flooded and non-flooded areas, a model was developed using the model-maker utility of Erdas Imagine (v. 10) to determine flooded areas. The derived flood map was then compared with the false color composite image and draped over the DEM. It was noticed that some misclassifications still existed, which originated from the wet soil category with sparse vegetation, such as open space. As increasing soil moisture leads to a drop in soil albedo during a flood, non-flooded pixels such as parks could have reflectance values similar to flooded pixels (Sheng et al. 1998). As a result, isolating purely flooded pixels from non-flooded pixels might be challenging. As these comprised few areas in the image, they had to be removed from the final model by digitizing with the AOI tool and then masking them out. A rerun of the model reasonably improved the result obtained earlier. Pixel values of >63 for non-water and ≤63 for floods were used to distinguish land and water boundaries in the TM bands (4+7). Finally, a binary raster was generated, showing only flooded and non-flooded areas.

Deriving only the flooded pixels from SAR, particularly in urban areas, is very exigent, as an urban area constitutes various man-made features that act as strong reflectors (Solbø and Solheim 2004). Despite this, a number of techniques have been used that encompass supervised, unsupervised, automated, semiautomated, and thresholding to delineate floods from radar data in urban environments (Mason et al. 2007, 2010; McMillan et al. 2006; Dewan et al. 2006, 2005; Solbø et al. 2004; Solbø and Solheim 2004). A simple model is that flooded areas are much smoother than surrounding dry lands and behave as specular reflectors, resulting in low backscatter. Thus, spectral information in terms of low backscatter value may be useful to distinguish flooded and non-flooded areas. A threshold value is then determined, and intensities that are less than the cutoff value are considered flood, and those above the threshold value are regarded as dry land.

Many studies have employed this technique to map flooded versus non-flooded areas (see Chap. 2). For example, using a density-slicing and threshold technique, Imhoff et al. (1987) demonstrated that SAR provided superior results compared to Landsat image in the identification of flooded areas, revealing the efficacy of this method. However, the cutoff value to separate flooded and non-flooded areas may depend on various factors, particularly in urban environments. For example, wind-induced ripples may enhance the backscatter of water surfaces, which could introduce difficulties in deciding the optimum threshold value for flooded land (Solbø and Solheim 2004; Liu et al. 2002). In addition, turbid water and emergent vegetation can significantly enhance radar backscatter (Smith 1997).

Generally, backscatter values in terms of dB are calibrated from the digital number (DN) of the radar image and are subsequently used to determine flooded and non-flooded areas. However, the calculation of dB requires a number of readily available parameters such as incidence angle and gain constants. As most of the data used are historical, only one had the required information to convert the image DN to dB.

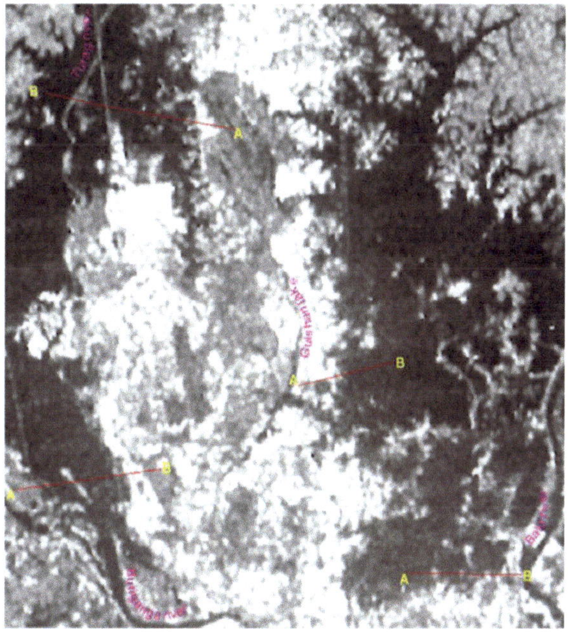

Fig. 4.2 Location of land–water threshold profiles (not to scale)

Hence, it was not possible to convert DN to dB. The original DN was therefore used to classify each radar image into discrete land-cover categories, that is, flooded and non-flooded pixels. A modified version of the method by Liu et al. (2002) was adopted and used to estimate the flood extent from SAR; however, it differs in working scale.

First, a number of profiles from land–water and water–land were drawn (see Fig. 4.2) interactively using the spatial profile tool on each image, where the x-axis represented distance and the y-axis characterized pixel value in terms of the intensity of the radar echo (see Fig. 4.3). Second, a cutoff value for land–flood was decided by averaging the values of water pixels and their neighboring pixels. Based on the optimal threshold value for each image, a rule-based approach was adopted (Wang et al. 2002) to extract flood boundaries from SAR. The following rule was applied to all SAR data using the model-maker utility of the image-processing software.

{DN<X then pixels represented “water”}{DN>=X then pixels represented “non-water/land”} where X represents provided threshold value

To separate flooded and non-flooded areas, different cutoff values were tested. If a pixel’s DN value was satisfactory, it was assigned to the water category; otherwise, non-flooded. Although the selection of the threshold values may appear to be arbitrary, the optimal threshold method suggested by Fung and LeDrew (1988) was used to substantiate the result. While this technique assisted in the determination of the optimal threshold value and the subsequent classification of radar data, the misclassification of land area as flooded pixels was still found in the classified image. For example, airport runways and parks behaved as specular reflectors, which resulted

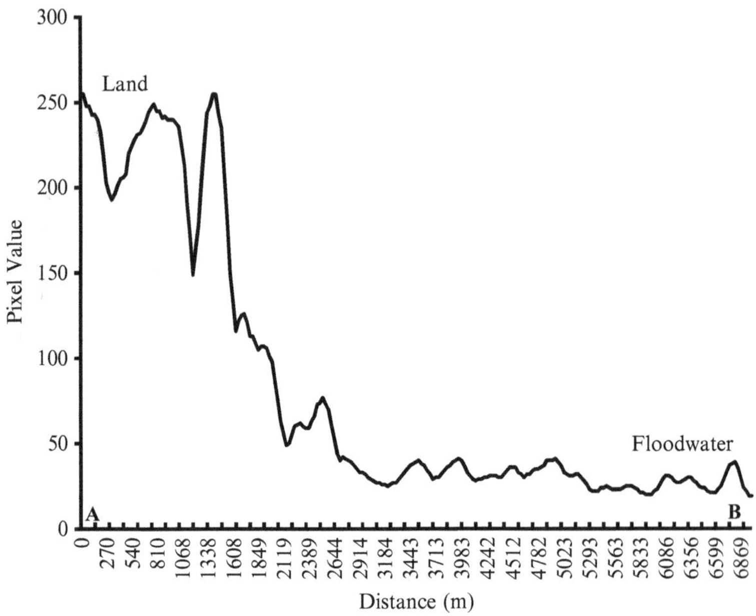

Fig. 4.3 Land–water profile for determining the threshold value

in low backscatter values. Likewise, boats and steamers on the rivers produced higher radar returns. These man-made features were digitized using AOI tools and masked out from the final model. A binary raster was then obtained for each year where only two classes existed: flooded and non-flooded.

In Dhaka, flood maps are only available for major events at a small scale. As this study used normal flood-time images in addition to major floods, the flood depth map was therefore calculated to support not only the evaluation of satellite-derived flood maps, but also the estimation of flood damage according to different land-use category. The flood depth was calculated by developing a model in the model builder of ArcGIS (v. 10). First, a watershed boundary of five major rivers in the study area was mapped using the DEM and the Arc Hydro tool (Maidment 2002). A point file representing gauge stations was then included with the catchment feature data, and HWL information was attached as an attribute table for each year. A water surface map was subsequently generated using the HWL for each year. Using the raster calculator, land elevation was subtracted from each of the water surface maps on pixel by pixel (de Brouder 1994), and new data layers were generated depicting the depth of flooding for each flood event. The resulting flood depth maps can further be classified to display various flood depths such as shallow, moderate, and deep flooding (see Fig. 4.4). To evaluate the results of the flood depth maps developed here, only the peak flood depth maps of 1988 and 1998 were collected (GoB 2000), as depth maps for other years were not available. It should be noted that both flood depth maps differed in spatial extent and the method of estimation; hence, only the

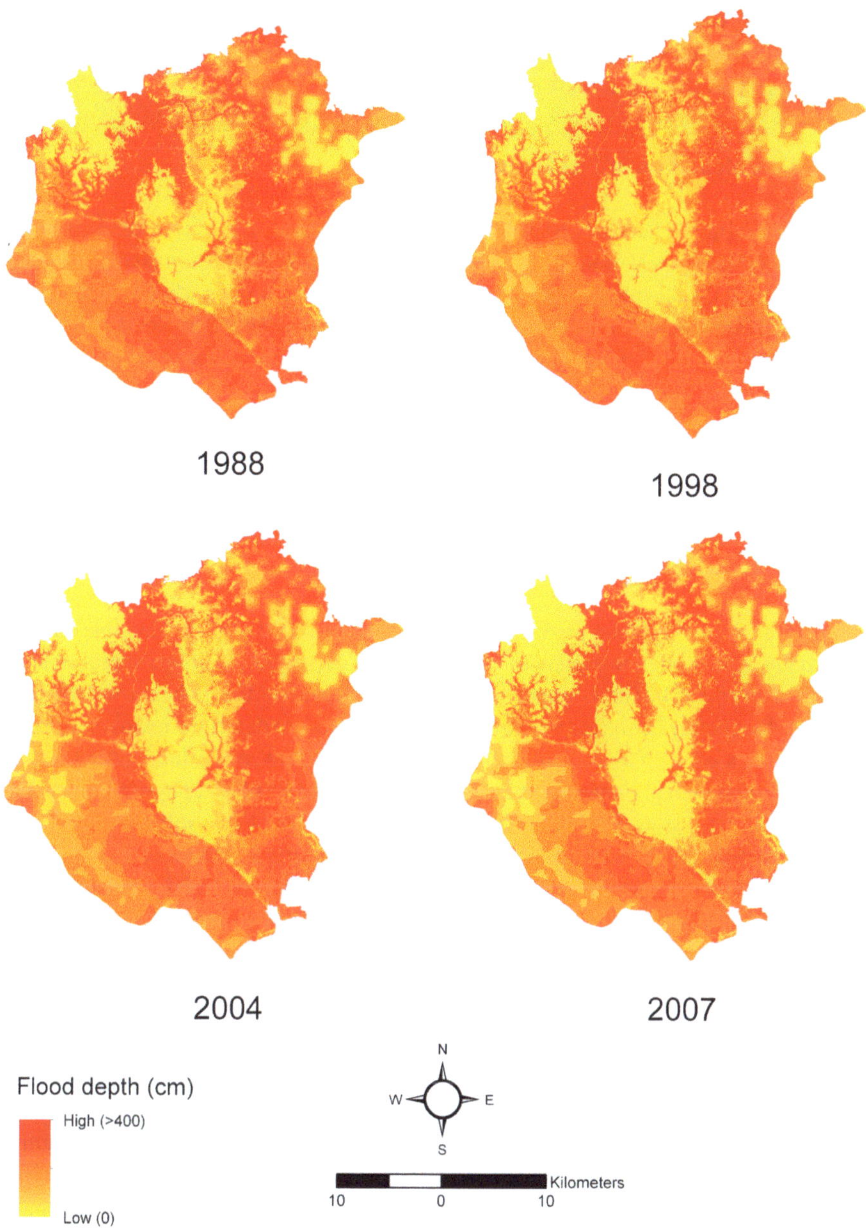

Fig. 4.4 Flood depth maps for 1988, 1998, 2004, and 2007

areal extents present in both datasets were evaluated through swiping, blending, and visual interpretation. The results showed a very good spatial agreement between GoB maps and the depth maps developed here.

4.2.2.4 Calculation of Net Inundated Areas

Upon extracting the flooded and non-flooded pixels for each year, every classified image was then superimposed onto dry seasonal water bodies in order to estimate the net inundated areas. It is necessary to note here that the dry seasonal water bodies were extracted from the 2000 land-use/land-cover map of the study area. Thus, the flooded area percentage for each flood-time image can be calculated using the following equation:

$$\text{Inundation area percentage}\left(\%\right) = \frac{a}{a+b} \times 100 \tag{4.1}$$

where a = inundated area and b = non-flooded area.

4.2.2.5 Flood Damage Assessment

Flood damage can be categorized into tangible and intangible losses. Tangible damage is usually expressed in monetary terms and heavily depends on damage appraisal techniques (Parker 2000). This is usually calculated according to different land-use categories and is subsequently converted into financial terms in order to depict the losses of different land-use classes such as commercial and residential. Conversely, intangible loss can rarely be expressed in monetary terms (Parker 2000), as loss of life, physical injury, and loss of historical sites are impossible to measure with monetary value. A number of factors such as flood duration, water depth, flow velocity, sediment size, and rate of water-level rises during the flood onset influence the effects of floods (van der Sande et al. 2003). However, flood depth and flood extent are two important factors that contribute significantly to direct flood losses (Green et al. 1994).

Two approaches are used to estimate flood damage in an urban area: parcel-based and grid-based (Su et al. 2005). The parcel-based approach requires a great deal of information for each land-use category, such as type of housing, dwelling type, and number of factories, where flood loss is a function of the estimated floodwater depth (Boyle et al. 1998). The grid-based approach divides the region into grid cells of equal areas. Information from the census is then aggregated and assigned to each cell by assuming that socioeconomic activities are homogenous within the cell (Su et al. 2005).

The widely used concept of the stage–damage curve (Su et al. 2005; Dutta et al. 2003, 2005; Smith 1994; White 1964) was used to estimate flood loss for housing and road networks, as other land-use categories were not available. One important shortcoming of this method is that water depth and building damage only explain part of the data variance (Merz et al. 2004). As working with extensive data at the community level is problematic in a developing country such as Bangladesh, this study estimated flood damage using generalized values. To estimate flood loss in each land-use category, the economic value of the respective land use must be

Table 4.2 Flood damage function for housing types and major roads

Water depth (m)	*Katcha*[a]	*Semi-pucca*[a]	*Pucca*[a]	Road[a]
0	0	0	0	0
1	0.7	0.4	0.2	0.2
2	1.0	0.7	0.5	0.8
3	1.0	1.0	0.9	1.0
4	1.0	1.0	1.0	1.0
5	1.0	1.0	1.0	1.0
6	1.0	1.0	1.0	1.0
7	1.0	1.0	1.0	1.0

[a]Repairing costs for *katcha*, *semi-pucca*, and *pucca* were 50, 180, and 250 (1,000) BDT, respectively, for the 2004 flood. Similarly, the repairing cost for each housing type was 100, 250, and 380 (1,000) BDT, respectively, for the 2007 flood. For road damage estimation, the repairing cost was 2.24 and 3.18 (100,000) BDT per km

known for the year under consideration. The author derived flood loss values from literature and field visits in terms of repairing costs (Rahman 2004; Alam and Zakaria 2002; Chowdhury and Sato 2000; FAP 8A 1991). The depth–damage curve for housing types was only developed for the 2004 and 2007 floods. Likewise, road damage was estimated for the 1998 and 2007 floods, and the monetary value to derive flood–damage curves are lacking for other years; therefore, estimating flood damage for every year is not viable to characterize the effects of floods. It should be noted that the author has attempted to obtain stage–damage curves for three types of housing (*katcha*, *semi-pucca*, and *pucca*) and for the major roads, assuming that the repairing cost was uniform throughout the study area.

As Dhaka is growing quickly, the flood damage derived here may not fully represent the actual flood loss; however, the estimated damage is intuitive to discern the effect of flooding. Note that the original road data obtained from the DAP was used to estimate the 1998 flood loss, while the updated road data were employed to obtain infrastructural damage for the 2007 flood (see Sect. 4.2.1). The flood loss was only related to water depth; hence, other parameters influencing flood damage were not considered. To obtain the depth–damage curve, flood depth maps were intersected with housing and road data, and the number of houses in each flood depth category was calculated. Similarly, the length of major roads under water was determined. The flood loss was determined using the weighted repairing cost for each housing type and major roads. A damage function (van der Sande et al. 2003) was then decided by linearly interpolating damage statistics on a scale of 0–1, where 0 means no damage and 1 represents complete loss (see Table 4.2). The flood loss was estimated using the formula below:

$$D = h * w * n \tag{4.2}$$

where D is the flood damage, h represents the repairing cost of each housing subtype/road, w means damage proportion of water depth, and n denotes the number of houses/length of major roads in each flood depth category.

4.3 Assessing the Accuracy of Flood Maps

Generally, the classification accuracy of RS data refers to the comparison of two datasets—one based on the analysis of image data and the other comprising reference data obtained from the field or via other sources such as historical maps (Congalton 1991). In assessing the classified maps from the satellite data, the most widely used approach is the confusion or error matrix (Rosenfield and Fitzpatrick-Lins 1986). The number of corrected pixels in each class and those incorrectly assigned to other classes are arranged in the error matrix. It has been suggested that at least 50 randomly selected pixels of each land-cover category should be checked against reference data (Hay 1979). A stratified random sampling scheme is recommended to ensure that an approximately equivalent number of samples are selected for each class (Congalton 1991). To assess the accuracy of maps extracted from satellite data, 100 stratified random pixels—ensuring 50 pixels for each category—were first generated. Then, using the historical flood maps and flood depth data, the sampling pixels were verified individually, and the results were derived in an error matrix. For each map, the overall accuracy was calculated as the number of correctly classified pixels divided by the total number of testing pixels (Story and Congalton 1986). A nonparametric kappa test was also used to measure the classification accuracy (Rosenfield and Fitzpatrick-Lins 1986) using the equation below:

$$k = \frac{N\sum_{i=1}^{r} x_{ii} - \sum_{i=1}^{r}\left(x_{i+}x_{+i}\right)}{N^2 - \sum_{i=1}^{r}\left(x_{i+}x_{+i}\right)} \tag{4.3}$$

where k is the kappa coefficient, r is the number of rows in the error matrix, x_{ii} is the number of observations in the major diagonal (row i and column i), x_{i+} is the total number of observations in row i, x_{+i} is the total number of observations in column i, and N is the total number of observations in the matrix.

4.4 Spatial and Temporal Distribution of Floods

The temporal dynamics of flood extent in different years (in terms of percent of inundation area) were computed and are shown in Table 4.3. Using Eq. 4.1, the net flooding area can be estimated. For example, the February 2000 data represent the dry season, while 16 October 1988 characterizes the flood period. Therefore, water areas in both dry and flood seasons (3.4%) were normal water bodies such as rivers, lakes, and ponds. The water areas in the flood season and non-water areas in the dry season (43.6%) represent inundated areas in the flood season. The non-water areas in both dry and flood seasons (52.5%) represent non-flooded areas (land area only). The water areas in the dry season and the non-water areas in the flood season (0.5%) show the classification error. Thus, the net flooded areas,

Table 4.3 Inundation area percentages in Dhaka, 1988–2009

	Dry season (February 2000)		
16 October 1988	Water	Non-water	Total
Water	3.44	43.57	47.01
Non-water	0.54	52.45	52.99
Total	**3.98**	**96.02**	**100.00**
	Dry season (February 2000)		
26 August 1998	Water	Non-water	Total
Water	2.99	42.49	45.47
Non-water	0.61	53.92	54.53
Total	**3.59**	**96.41**	**100.00**
	Dry season (February 2000)		
13 August 2000	Water	Non-water	Total
Water	2.79	27.78	30.56
Non-water	1.20	68.24	69.44
Total	**3.98**	**96.02**	**100.00**
	Dry season (February 2000)		
8 August 2001	Water	Non-water	Total
Water	2.56	25.43	27.99
Non-water	1.42	70.59	72.01
Total	**3.98**	**96.02**	**100.00**
	Dry season (February 2000)		
22 August 2003	Water	Non-water	Total
Water	2.79	25.88	28.68
Non-water	1.19	70.14	71.32
Total	**3.98**	**96.02**	**100.00**
	Dry season (February 2000)		
23 July 2004	Water	Non-water	Total
Water	2.69	35.89	38.57
Non-water	1.30	60.13	61.43
Total	**3.98**	**96.02**	**100.00**
	Dry season (February 2000)		
11 August 2005	Water	Non-water	Total
Water	2.48	25.84	28.32
Non-water	1.51	70.17	71.68
Total	**3.98**	**96.02**	**100.00**
	Dry season (February 2000)		
3 August 2007	Water	Non-water	Total
Water	2.99	37.01	40.01
Non-water	0.99	59.00	59.99
Total	**3.98**	**96.02**	**100.00**
	Dry season (February 2000)		
10 July 2009	Water	Non-water	Total
Water	2.65	25.24	27.89
Non-water	1.34	70.77	72.11
Total	**3.98**	**96.02**	**100.00**

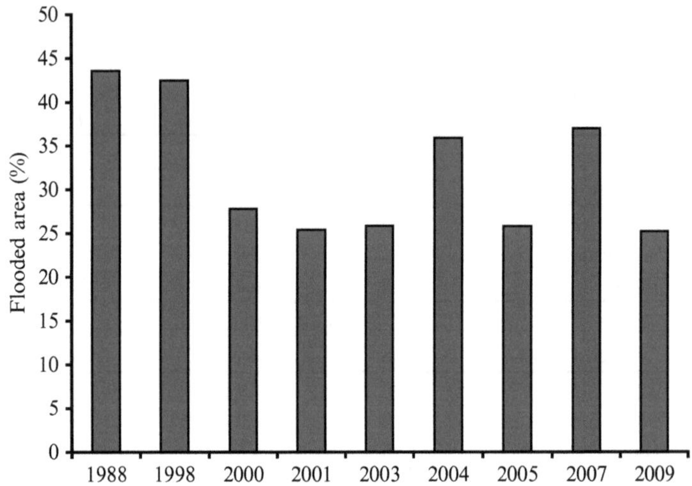

Fig. 4.5 Percentage of net inundated areas in Dhaka, 1988–2009

excluding normal water bodies, were 43.6%. This result resembles the statistics obtained by Sado and Islam (1997). Using a MOS–MESSR image of 17 October 1988, they reported that at least 41.1% of the areas were flooded during the 1988 flood. Two factors might have contributed to this discrepancy of around 2.5%. First, flood water is rapidly moving, and therefore an image of a later date might not have similar information as found in this study. Second, cloud cover in the MESSR image restricted to derive the actual flooding areas as concluded by their study. Nevertheless, the net inundation areas could be greater than this figure, as none of the studies could employ a peak flood-time image for the 1988 flood. Flood estimation for other years can therefore be explained in the same manner (see Table 4.3). For example, in 1998, the net flooded area was 42.5% of the total study area. It is interesting to note that flooding became pervasive in two other major events of 2004 and 2007. These two floods were moderate events, and the effects were relatively low in Dhaka compared to other parts of Bangladesh (Bala et al. 2009). However, as shown in Table 4.3, at least 35.9 and 37% of the areas were flooded in 2004 and 2007, respectively (see Fig. 4.5). In a normal year, flood-prone areas comprise at least 25% of Dhaka.

The spatial distribution of flooding for different years is shown in Fig. 4.6a, b. A close inspection of these maps revealed that flooding was widespread during 1988, while in 1998, most of the embanked areas were flood-free. However, pluvial flooding affected a considerable portion of the embanked areas (see Chap. 3). Similar patterns of inundation can be observed in other events. As expected, the spatial extent of flooding considerably increased in areas beyond the embankment due to changes in the hydrological regime linked with flood control works, which should have substantial effects on agricultural production and the livelihood of marginal people living in those areas.

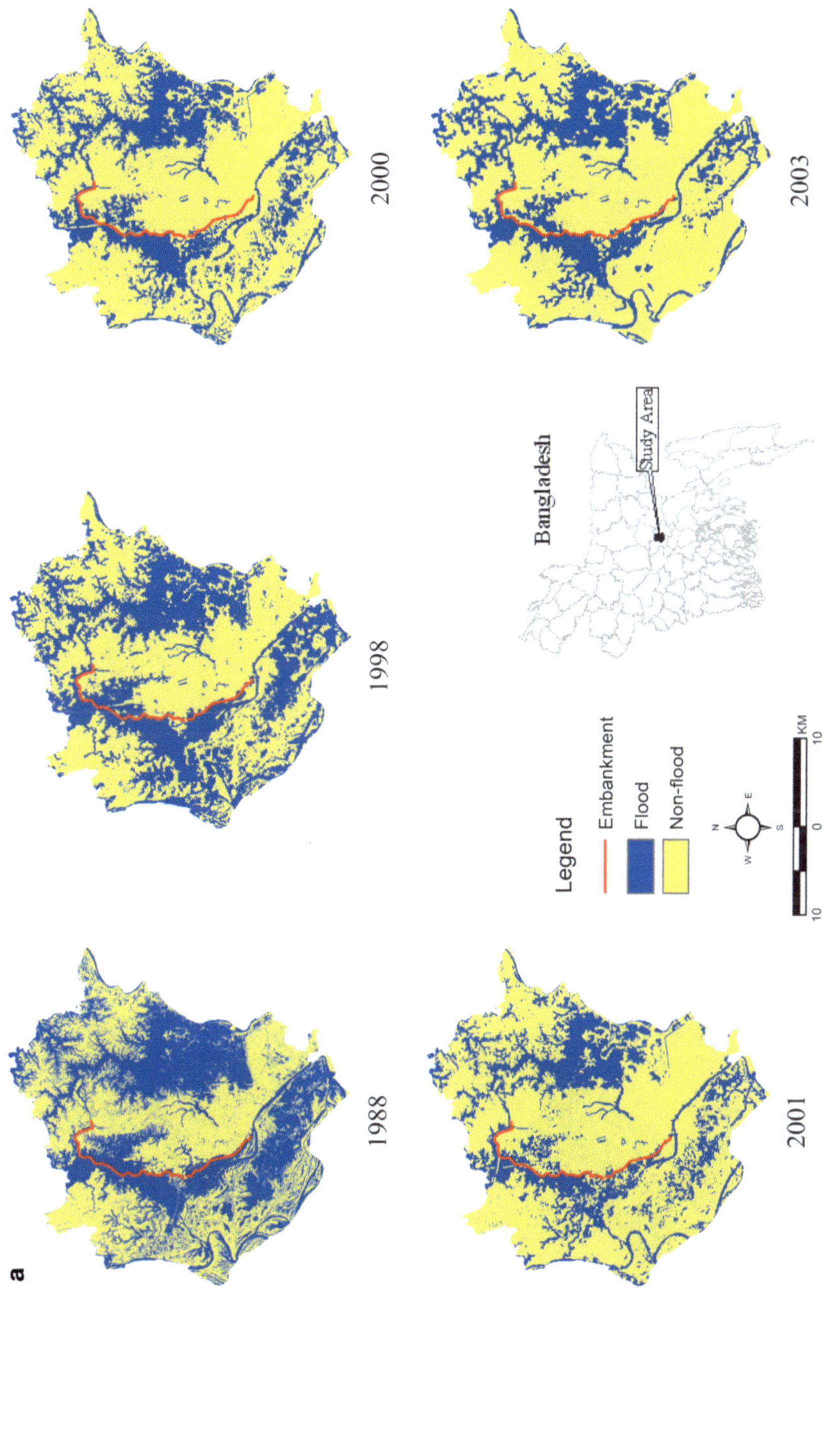

Fig. 4.6a Spatial patterns of flooding in Dhaka megacity, 1988–2003

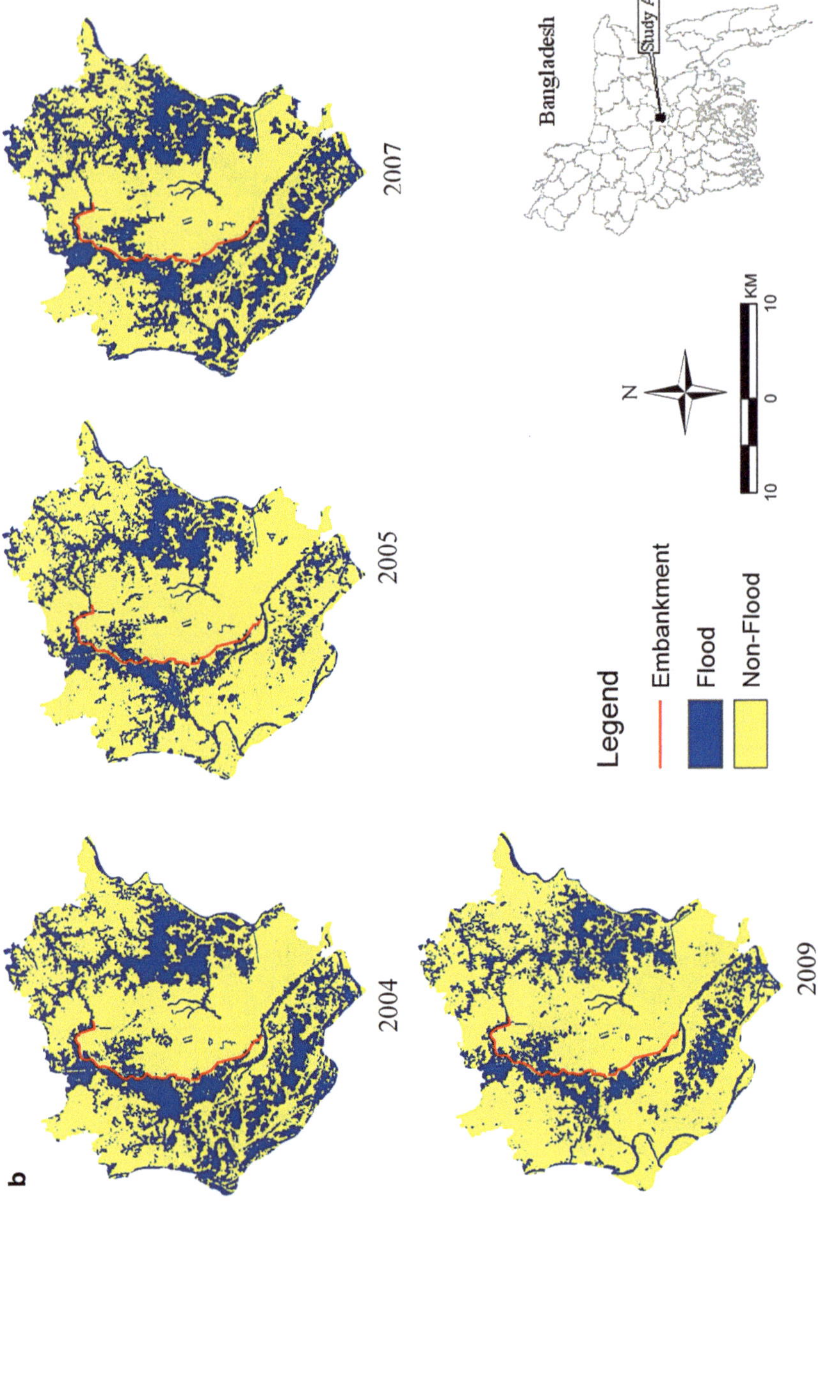

Fig. 4.6b Spatial patterns of flooding in Dhaka megacity, 2004–2009

4.5 Accuracy of Flood Maps

As mentioned above, the flood maps derived from the multi-temporal satellite data were subject to accuracy evaluation, which was conducted by considering the ancillary information obtained from the field in the form of historical flood extent maps and depth data generated in this study. It was found that the overall accuracy of the flood maps ranged between 68 and 75%. Overall accuracy and corresponding kappa statistics are shown in Table 4.4. The highest accuracy was obtained for the 1998 and 2004 flood maps, while the lowest accuracy was in 2001. A summary of the classification accuracies in terms of producers' and users' accuracy revealed that the producers' accuracy for the flood category ranged from 65 to 80%, while the users' accuracy was between 72 and 82%. In contrast, producers' accuracy for the non-flooded category was in the range of 69–79%, whereas the users' accuracy ranged from 60 to 70%. These results demonstrated that the flood maps obtained from satellite data agree with the observed information sourced from fields, historical maps, and flood depth data.

4.6 Flood Damage Estimation

The flood depth–damage curve for housing types is shown in Fig. 4.7a, b. Surprisingly, all of the housing categories exhibited higher damage, and the effects were higher in the depth between 2 and 6 m. As expected, flood damage was higher for *katcha* houses, but because of their low repair costs, the *pucca* and semi-*pucca* categories were still showing higher flood loss due to their high repair costs. Only the major roads in the study area were considered to construct the depth–damage curve, which is presented in Fig. 4.8. Interestingly, the damage was higher during the 2007 flood, even though the magnitude of that flood was considerably less than the 1998 flood. Two factors may have contributed to this: the persistent development of floodplains

Table 4.4 Overall accuracy and kappa coefficient of satellite-derived flood maps, 1988–2009

Year	Overall accuracy (%)	Kappa coefficient
1988	71.0	0.42
1998	75.0	0.50
2000	72.0	0.44
2001	68.0	0.36
2003	73.0	0.46
2004	75.0	0.50
2005	69.0	0.38
2007	73.0	0.46
2009	68.0	0.36

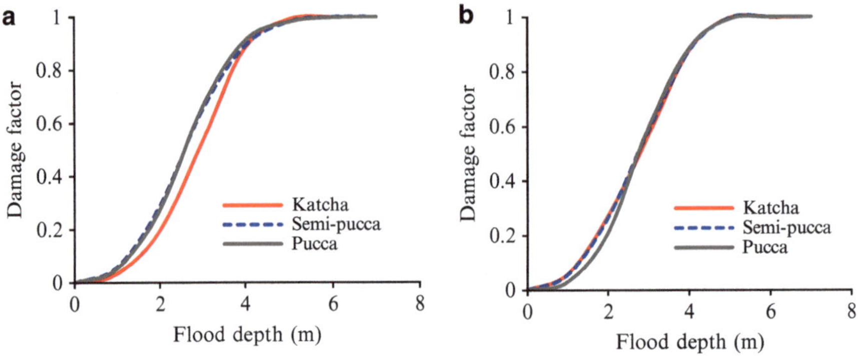

Fig. 4.7 Depth–damage curves for housing types (**a**) 2004 flood, (**b**) 2007 flood

Fig. 4.8 Depth–damage curves for major roads during 1998 and 2007 floods

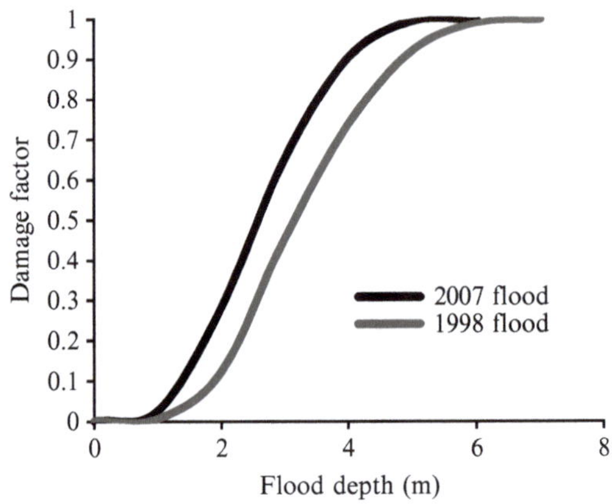

and the cost associated with repairing roads in a latter period. The value for repairing roads was certainly higher in 2007 than in 1998 because of the continuous devaluation of the Bangladeshi currency against the US$. This may have resulted in higher repairing costs in 2007, thereby resulting in higher damage. However, the information derived here is crucial, as roads are critical infrastructures that support relief and rescue operations during and after the flood. Figure 4.9 shows the spatial distribution of the estimated housing damage in the study area at community level. The aggregated financial damage maps revealed that flood loss is higher in the peripheral zones and areas that are prone to floods in monsoons. For example, census districts that are close to river networks predominantly with poor quality housing (e.g., *katcha* type) experienced high monetary loss during 2004 and 2007 floods. Kamrangir Char, Aminbazar, Kaundia, Kaliganj, Mirerbag, Pubgaon, etc. were within this group. In contrast, communities that are within embanked areas (e.g., central, north, and old Dhaka) experienced no or low financial damage. Interestingly, well-off neighborhoods such as Dhanmondi, Gulshan, and Banani suffered medium flood damage during 2004 flood (Fig. 4.9) which could be the result of pluvial flooding.

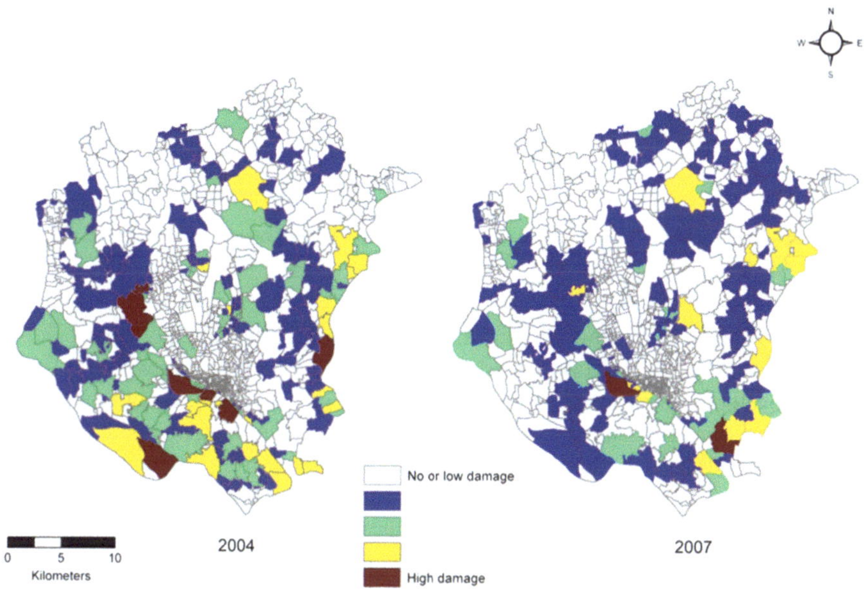

Fig. 4.9 Estimated financial damage maps based on housing types in the 2004 and 2007 floods, by community (See text for explanation)

4.7 Discussions

The estimation of flood extents using multi-temporal satellite data revealed that inundation was widespread during 1988 and 1998. However, the net inundation areas did not significantly differ between 1988 and 1998, even if a large number of resources were used to construct flood control works after the 1988 flood. Although the spilling of river waters was the main governing factor in the 1988 and 1998 floods, the magnitude of the 2004 and 2007 floods largely resulted from mismanagement. As a result, both pluvial and fluvial flooding were pervasive in 2004 and 2007 (Alam and Rabbani 2007). As Dhaka's drainage system has not been evolved in pace with rapid urban growth, the encroachment of natural water bodies along with a dilapidated drainage system is aggravating the situation of pluvial floods in particular, causing immense suffering.

Studies revealed that structural solutions provide a false sense of security that is often counterproductive (Few 2003; Green et al. 1991), and they also enhance the vulnerability of lives and property to natural hazards (Etkin 1999; Haque 1994). Although flood control works have provided some security to the embanked areas, the overall achievement of these structures is questionable. Consequently, bringing more areas under flood control requires more money for maintenance, which is difficult to bear for a developing country like Bangladesh. As flood structures encourage people to encroach on floodplains and low-lying lands, a 60-km embankment along

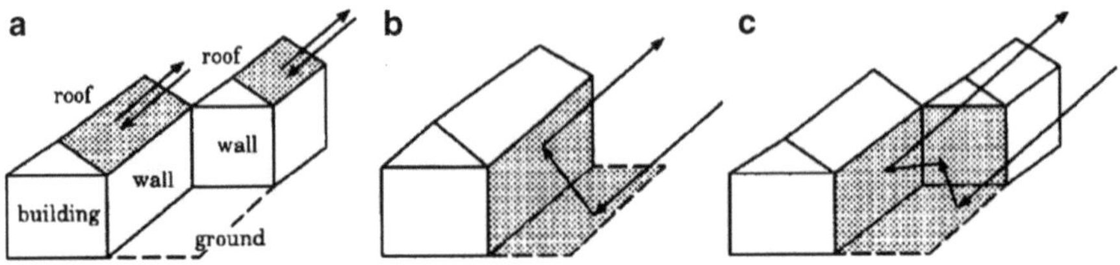

Fig. 4.10 Backscattering mechanisms in urban areas: (**a**) Single bounce scattering; (**b**) Double bounce scattering; (**c**) Triple bounce scattering (Adapted from Dong et al. 1997)

the Balu River would allow further development, which has already been evident since its announcement (Masood and Takeuchi 2012). Flood control works aid people in developing land for purposes other than agriculture; thus, Bangladesh must take into account the issue of food security for its ever-increasing population. As the effects of floods are apparent on agricultural crops, saving crops should be the country's priority. To ensure food security in the coming years, Bangladesh must minimize flood damage on agricultural crops; otherwise, increasing crop damage could lead to a severe shortage of food (Mirza 2011; Douglas 2009).

To manage floods effectively, human response systems to natural hazards should be incorporated into the current flood management practice, as this parameter is no less important in reducing flood risk (Krasovskaia et al. 2001). The city should formulate a flood management practice that not only saves lives and property from floods but also saves precious natural resources from rapid encroachment. As noted by Barua and van Ast (2011), current flood management practice is neither conducive for people and the environment nor does it support modern water governance. Therefore, the local government of Dhaka must mainstream flood hazards into development practices, particularly in urban planning, which could considerably alleviate its increasing vulnerability to flood.

The results of accuracy assessment varied from year to year. Even though radar data produced the highest overall accuracy for two years (1998 and 2004), the overall accuracy for other years was low among datasets. On the radar image, water and land areas can easily be separated due to their distinct tonal variation. Generally, water pixels appear in a smooth gray tone due to specular scattering, while land areas are often characterized by very bright tones, representing corner reflectors of urban features. Analysis of the error matrix showed that there was misclassification between water and non-water pixels. This misclassification could be attributed to several factors associated with radar images. For instance, it was difficult to spectrally separate open spaces (parks, institution grounds, golf courses with grassy characteristics, and linear features, i.e., roads and airport runways) because of their intermediate tonal characteristics. These groups of features generally appeared as medium gray, and their DN values were slightly higher than water pixels, making them difficult to isolate.

A difficulty was also encountered with the occurrence of low returns from features adjacent to water bodies, such as roads surrounded by lakes and vegetation. These are common problems in studying urban surfaces using radar data (Dong et al. 1997), as demonstrated in Fig. 4.10. Some of the map errors likely stemmed from interactions between ground features (land/water boundaries), surface roughness, moisture content, and radar parameter configurations (incidence angle, polarization, and wavelength). These are complex phenomena and are subject to thorough research (McMillan et al. 2006; Lee and Lunetta 1995; Ramsey 1999). Radarsat's polarization may have been another source of error. Sensors with HH polarization are known to be less sensitive to the changes in vegetation moisture content than cross-polarized sensors (HV and VH) (Avery and Berlin 1992). Although a study asserted that the higher incidence angle is relatively good for flood detection (Solbø and Solheim 2004), another study suggested that the larger incidence angle might give rise to specular reflectance for a given surface (Sokol et al. 2000). This may provide an additional explanation for confusion between the classes. Another source of error could be the misregistration of images (Townshend et al. 1992), despite the fact that minimal RMS errors were taken into account.

However, this study performed relatively better in mapping fluvial flooding in an urban landscape compared to other SAR and TM studies. Radar data were used to study urban flooding (Mason et al. 2010; McMillan et al. 2006; Dewan et al. 2006, 2005), forest flooding (Hess et al. 1995; Townsend 2001), tidal flooding (Ramsey 1995), and the detection of freshwater wetland flooding (Töyora et al. 2002; Adam et al. 1998). Among these studies, Townsend (2001) reported very high accuracy (93.5%), and others achieved variable accuracy. For instance, Töyora et al. (2002) used an image fusion approach to study wetland flooding. Their result demonstrated that a SAR image on its own produces 70% accuracy; however, when it is used with optical data (SPOT), the accuracy increased to 92%. Adam et al. (1998) also achieved over 90% mapping accuracy for flood detection in freshwater wetlands using Radarsat data. In contrast, Henderson et al. (1998) conducted a study using SAR data from Radarsat and ERS sensors in a complex urban setting and reported extremely poor accuracy for both sensors. Compared to the result obtained by Henderson et al. (1998), the results of this study are encouraging, as overall accuracy is significantly higher than in their study. The accuracy of TM mapping was also higher than the study by Imhoff et al. (1987).

Unlike developed countries, the scarcity of data has been restricted to the development of detailed depth–damage curves for each land-use category. It should be noted that this study used average repair costs for houses in three categories, as detailed information on dwelling types such as single- and multifamily housing were not available. Hence, the estimated damage needs to be interpreted carefully, as repair costs might have been overestimated because people and respective government agencies tend to apply higher values for repairing structures (Chowdhury and Sato 2000). Similarly, the damage function for roads only accounted for major roads, which may not be representative of other categories such as regional highways and feeder roads. However, to the author's knowledge, the depth–damage curve developed in this study is the first attempt to estimate flood loss according to water depths in Dhaka. Hence, this information may be of significant assistance in developing efficient flood preparedness and mitigation programs to combat future flood-related effects.

4.8 Summary

This chapter presented flood mapping during 1988–2009 using multi-temporal RS data. Both optical and microwave sensors were used to delineate the spatial and temporal distribution of floods in the megacity of Dhaka. Flood extents were estimated by digitally classifying satellite data with a threshold algorithm. It was found that Dhaka was heavily inundated during the 1988 and 1998 floods. While surface water flooding was pervasive during these two events, the floods in 2004 and 2007 were the result of the combined effects of both pluvial and fluvial floods. In a normal period, at least 25% of Dhaka is subject to inundation; in an abnormal event, this figure could rise to more than 40%. Although flood control works are supporting some parts of Dhaka, pluvial flood as a result of dilapidated drainage systems, along with the encroachment of floodplains, is elevating flood risk potential with the progress of time. The accuracy of flood maps showed a good agreement between digitally classified maps and reference data. Even though there is some confusion between land-cover categories that inhibit superior results, the multi-temporal flood maps developed here could be used to estimate hazards and risks from floods in order to develop effective flood management systems. Flood depth–damage curves for housing types and major roads showed that flood loss in the study area is rising despite major flood control initiatives, revealing that existing policies for flood hazard abatement need to be reassessed, as increasing precipitation is expected in the future. In the absence of historical records, which are common in developing countries, particularly Bangladesh, satellite data is an excellent alternative to generate baseline information and thus support the estimation of likely patterns of risk from natural hazards such as floods.

References

Adam S, Wiebe J, Collins M, Pietroniro A (1998) RADARSAT flood mapping in the Peace-Athabasca Delta, Canada. Can J Remote Sens 24(1):69–79

Alam M, Rabbani MG (2007) Vulnerabilities and responses to climate change for Dhaka. Environ Urban 19(1):81–97

Alam MJB, Zakaria M (2002) Design and construction of roads in flood affected areas. In: Ali MA, Seraj SM, Ahmad E (eds) Engineering concerns of flood. Bangladesh University of Engineering and Technology, Dhaka, pp 91–100

Avery TE, Berlin GE (1992) Fundamentals of remote sensing and airphoto interpretation. Macmillan, New York

Bala SK, Islam AKMS, Chowdhury JU, Rahman MR, Haque MA, Khan MSA, Salehin M (2009) Performance of flood control works around Dhaka city during major floods in Bangladesh. A paper presented in the 2nd international conference on water and flood management, Dhaka

Bangladesh Bureau of Statistics (BBS) (2003) Population census – 2001. Community series – Dhaka, Narayanganj, Gazipur. Bangladesh Bureau of Statistics, Ministry of Planning, Dhaka

Barua S, van Ast JA (2011) Towards interactive flood management in Dhaka, Bangladesh. Water Policy 13(5):693–716

Baumann PR (1999) http://www.research.umbc.edu/tbenja1/baumann/mode2.html. Accessed 12 Jan, 2004

Bhavasar PD (1984) Review of remote sensing applications in hydrology and water resources management in India. Adv Space Res 4(11):193–200
Boyle SJ, Tsanis IK, Kanaroglou PS (1998) Developing geographic information system for land use impact assessment in flooding condition. J Water Resour Plan Manage 124(2):89–98
Chowdhury MR, Sato T (2000) The impact of household characteristics on flood damage – a case study. J Floodplain Manage 1(2):35–56
Chowdhury JU, Rahman R, Bala SK, Islam AKMS (1998) Impact of 1998 flood on Dhaka City and performance of flood control works. Institute of Flood Control and Drainage Research, Bangladesh University of Engineering and Technology, Dhaka
Congalton RG (1991) A review of assessing the accuracy of classifications of remotely sensed data. Remote Sens Environ 37(1):35–46
Consuegra D, Joerin F, Vitalini F (1995) Flood delineation and impact assessment in agricultural land using GIS technology. In: Carrara A, Guzzetti F (eds) Geographical information systems in assessing natural hazards. Kluwer Academic Publishers, Dordrecht, pp 177–198
Corbley KP (1993) Remote sensing and GIS provide rapid response for flood relief. Earth Observ Mag 2:28–30
de Brouder JAM (1994) Flood study in the Meghna – Dhonagoda polder, Bangladesh. In: Proceedings of the 15th Asian conference on remote sensing, Bangalore, India, 17–23 November
Dewan AM, Corner RJ (2012) The impact of land use and land cover changes on land surface temperature in a rapidly urbanizing megacity. In: IGARSS 2012, Munich, Germany
Dewan AM, Yeboah KK, Nishigaki M (2005) Flood mapping and damage evaluation in Greater Dhaka, Bangladesh with remote sensing. Asian Profile 33(5):495–512
Dewan AM, Yeboah KK, Nishigaki M (2006) Using synthetic aperture radar (SAR) data for mapping river water flooding in an urban landscape: a case study of Greater Dhaka, Bangladesh. J Hydrol Water Resour 19(1):44–54
Dong Y, Forster B, Ticehurst C (1997) Radar backscatter analysis for urban environments. Int J Remote Sens 18(6):1351–1364
Douglas I (2009) Climate change, flooding and food security in South Asia. Food Secur 1(2):127–136
Durand JM, Gimonet BJ, Debros JR (1987) SAR data filtering for classification. IEEE Trans Geosci Remote Sens GE-25(5):629–637
Dutta D, Herath S, Musiake K (2003) A mathematical model for flood loss estimation. J Hydrol 277(1–2):24–49
Dutta D, Khatun F, Herath S (2005) Analysis of flood vulnerability of urban buildings and population in Hanoi, Vietnam. Seisan Kenkyo 57(4):150–154
Etkin D (1999) Risk transference and related trends: driving forces towards more mega-disasters. Environ Hazard 1(2):69–75
Few R (2003) Flooding, vulnerability and coping strategies: local responses to a global threat. Prog Dev 3(1):43–58
Flood Action Plan (FAP) 8A (1991) Master plan study for Greater Dhaka protection project. Japan International Cooperation Agency, Dhaka
Flood Action Plan (FAP) 19 (1995) Satellite based radar for flood mapping in Bangladesh. ISPAN/ USAID, Dhaka
Fung T, LeDrew E (1988) The determination of optimal threshold levels for change detection using various accuracy indices. Photogramm Eng Remote Sens 54(10):1449–1454
Gillespie TW, Chu J, Elizabeth F, Thomas D (2007) Assessment and prediction of natural hazards from satellite imagery. Prog Phys Geogr 31(5):459–470
Government of Bangladesh (GOB) (2000) Baseline information study of the Dhaka combined flood control cum eastern bypass road project. Final report, Dhaka, p 106
Green CH, Tunstall SM, Fordham M (1991) The risks from flooding: which risk and whose perception. Disasters 15(3):227–236
Green C, van der Veen A, Wierstra E, Penning-Rowsell El (1994) Vulnerability refined: analysing flood impacts. In: Penning-Rowsell EC, Fordham M (eds) Flood across Europe: flood hazard assessment, modelling and management. Middlesex University Press, London, pp 32–68

Haque CE (1994) Flood prevention and mitigation in Bangladesh: the need for sustainable floodplain development. In: Goodland R, Edmundson V (eds) Environmental impact assessment and development. The World Bank, Washington, DC, pp 101–112

Hay AM (1979) Sampling designs to test land-use map accuracy. Photogramm Eng Remote Sens 45(4):529–533

Henderson FM, Lewis AJ (eds) (1998) Principles and applications of imaging radar. Wiley, New York

Henderson FM, Chasan R, Hart TF Jr, Portolese J (1998) Variations in SAR signal response from coastal wetlands in a complex urban setting. In: Proceedings of the 2nd international workshop on retrieval of bio- and geo-physical parameters from SAR data for land applications, ESTEC, Noordwijk, The Netherlands, 21–23 October, pp 487–492

Hess LL, Melack JM, Filoso S, Wang Y (1995) Delineation of inundated area and vegetation along the Amazon floodplain with SIR-C synthetic aperture radar. IEEE Trans Geosci Remote Sens 33(4):896–903

Horritt MS, Mason DC, Luckman AJ (2001) Flood boundary delineation from synthetic aperture radar imagery using a statistical active contour model. Int J Remote Sens 22(13):2489–2507

Imhoff ML, Vermillion C, Story MH, Choudhury AM, Gafoor A, Polcyn F (1987) Monsoon flood boundary delineation and damage assessment using space borne imaging radar and Landsat data. Photogramm Eng Remote Sens 53(4):405–413

Jensen JR (1979) Spectral and textural features to classify elusive land cover at the urban fringe. Prof Geogr 31(4):400–409

Krasovskaia I, Gottschalk L, Saelthun NR, Berg H (2001) Perception of the risk of flooding: the case of the 1995 flood in Norway. Hydrol Sci 46(6):855–868

Lee KH, Lunetta RS (1995) Wetlands detection methods. In: Lyon JG, McCarthy J (eds) Wetlands and environmental applications of GIS. CRC Press/Lewis Publishers, Boca Raton, pp 249–283

Liu Z, Huang F, Liu L, Wan E (2002) Dynamic monitoring and damage evaluation of flood in north-west Jilin with remote sensing. Int J Remote Sens 23(18):3669–3679

Maidment DR (2002) Arc hydro: GIS for water resources. ESRI Press, Redlands

Marco JB (1992) Flood risk mapping. In: Rossi G, Harmancioglu N, Yevjevich V (eds) Coping with floods. Kluwer Academic Publishers, Dordrecht, pp 353–373

Martin TC, Haque I, Hassan A, Werle D, Titley B (1996) Multi-temporal satellite radar imagery for flood monitoring in Bangladesh. In: Proceedings of 26th international seminar on remote sensing of environment, Vancouver, Canada, 25–29 March 1996, pp 353–358

Mason DC, Horritt MS, Dall' Amico JT, Scott TR, Bates PD (2007) Improving river flood extent delineation from synthetic aperture radar using airborne laser altimetry. IEEE Trans Geosci Remote Sens 45(12):3932–3943

Mason DC, Speck R, Devereus B, Schumann GJ-P, Neal JC, Bates PD (2010) Flood detection in urban areas using TerraSAR-X. IEEE Trans Geosci Remote Sens 48(2):882–894

Masood M, Takeuchi K (2012) Assessment of flood hazard, vulnerability and risk of mid-eastern Dhaka using DEM and 1D hydrodynamic model. Nat Hazard. doi:10.1007/s11069-011-0060-x

Mcmillan A, Morley JG, Adams BJ, Chesworth S (2006) Identifying optimal SAR imagery specifications for urban flood monitoring: a hurricane Katrina case study. Paper presented at 4th international workshop on remote sensing for post-disaster response, Magdalene College, University of Cambridge, Cambridge, UK, 25–26 September

Merz B, Kreibich H, Thieken A, Schmidtke R (2004) Estimation uncertainty of direct monetary flood damage to buildings. Nat Hazard Earth Syst Sci 4(1):153–163

Mirza MMQ (2011) Climate change, flooding in South Asia and implications. Reg Environ Chang 1(supp):95–107

Nuesch DR (1982) Classification of SAR imagery from an agricultural region using digital texture analysis. Int Soc Photogramm Remote Sens Int Arch 24-VII(1):231–240

Oliver CJ (1991) Information from SAR images. J Phys D: Appl Phys 24:1493–1514

Parker DJ (2000) Floods, vol 1. Routledge, London

Rahman AKM (2004) The impact of floods on major highways and mitigation options. In: Siddiqui KU, Hossain ANHA (eds) Options for flood risk and damage reduction in Bangladesh. The University Press Limited, Dhaka, pp 205–212

Ramsey EW III (1995) Monitoring flooding in coastal wetlands by using radar imagery and ground based measurements. Int J Remote Sens 16(13):2495–2502
Ramsey EW III (1999) Radar remote sensing of wetlands. In: Lunetta RS, Elvidge CD (eds) Remote sensing change detection. Ann Harbor Press, Chelsea, pp 211–243
Rosenfield GH, Fitzpatrick-Lins K (1986) A coefficient of agreement as a measure of thematic classification accuracy. Photogramm Eng Remote Sens 52(2):223–227
Sado K, Islam MM (1997) Satellite remote sensing data analysis for flooded area and weather study-case study of Dhaka City, Bangladesh. Ann J Hydraul Eng 41:945–950
Sheng Y, Su Y, Xiao Q (1998) Challenging the could-contamination problem in flood monitoring with NOAA/AVHRR imagery. Photogramm Eng Remote Sens 64(3):191–198
Smith DI (1994) Flood damage estimation – a review of urban stage-damage curves and loss functions. Water SA 20(3):231–238
Smith LC (1997) Satellite remote sensing of river inundated area, stage and discharge: a review. Hydrol Process 11(10):1427–1439
Sokol J, Pultz TJ, Bulzgis V (2000) Monitoring wetland hydrology in Atlantic Canada using multitemporal and multi-beam Radarsat data. In: Remote sensing and hydrology. International Association of Hydrological Sciences, Santa Fe, pp 536–538
Solbø S, Solheim I (2004) Towards operational flood mapping with satellite SAR. In: Envisat/ERS symposium, Salzburg
Solbø S, Pettinato S, Paloscia S, Santi E, Brusotti P, Solheim I (2004) Mapping of flooding in the Alessandria area with ERS. In: Proceedings of IGARSS 2004, pp 4689–4692
Story M, Congalton RG (1986) Accuracy assessment: a user's perspective. Photogramm Eng Remote Sens 52(3):397–399
Su M-D, Kang J-L, Chang L-F, Chen AS (2005) A grid-based GIS approach to regional flood damage assessment. J Mar Sci Technol 13(3):184–192
Townsend PA (2001) Mapping seasonal flooding in forested wetlands using multi-temporal RADARSAT SAR. Photogramm Eng Remote Sens 67(7):857–864
Townshend JRG, Justice CO, Gurney C, Mcmanus J (1992) The impact of misregistration on change detection. IEEE Trans Geosci Remote Sens 30:1054–1060
Töyora J, Pietroniro A, Martz LW (2002) A multi-sensor approach to wetland flood monitoring. Hydrol Process 16(8):1569–1581
van der Sande CJ, de Jong SM, de Roo APJ (2003) A segmentation and classification of IKONOS-2 imagery for land cover mapping to assist flood risk and flood damage assessment. Int J Appl Earth Observ Geoinf 4(3):217–229
Walker G, Burningham K (2011) Flood risk, vulnerability and environmental justice: evidence and evaluation of inequality in a UK context. Crit Soc Policy 31(2):216–240
Wang Y, Colby JD, Mulcahy KA (2002) An efficient method for mapping flood extent in a coastal floodplain using Landsat TM and DEM data. Int J Remote Sens 23(18):3681–3696
White GF (1964) Choice of adjustments to floods. Research paper no. 93. Department of Geography, University of Chicago, Chicago, IL
Wind HG, Nierop TM, de Blois CJ, de Kok JL (1999) Analysis of flood damages from the 1993 and 1995 Meuse floods. Water Resour Res 35(11):3459–3465

Chapter 5
Modeling Flood Hazards

Abstract Satellite-derived flood maps from 1988 to 2009 were used to construct a flood-affected frequency map of the study area. Similarly, a series of flood depth maps were used to generate a flood damage map for flood depth data. A flood hazard map was then developed by considering both the flood-affected frequency and flood depth data, simultaneously using a 2-D ranking matrix. The final flood hazard map was then combined with population and housing data to calculate the flood exposure for these two elements. The analysis revealed that 23% of the population is located in low- to high hazard zones, while 25% of human settlements are located in different flood hazard zones.

5.1 Introduction

Although there is a sharp difference between the manifestation of hazards and disasters, disaster research currently receives more attention than it did in the past. While hazard studies examine the beneficial and adverse effects of extreme events on society, disaster research focuses more on causes, physical risks, and emergency responses (White et al. 2001). Therefore, similar to other key terminologies of hazard literature, the definition of hazard varies. Nevertheless, identifying potential hazard areas is an integral part of the flood risk assessment process. Considerable efforts have been made worldwide to assess flood hazards using geospatial techniques (see Chap. 2). The primary intent of such efforts is to delineate hazardous areas, which can then assist in developing appropriate measures to lessen the negative effects of floods on people and the economy. A coarse- to moderate-resolution flood hazard map may have little relevance in developed countries; however, in developing countries, where precise information is seriously lacking, such hazard maps are extremely useful for saving the lives and property of millions of people, particularly marginal groups.

In this chapter, flood hazard assessment is presented by the integrated use of GIS and RS. To facilitate flood hazard analysis, flood frequency and floodwater

A.M. Dewan, *Floods in a Megacity: Geospatial Techniques in Assessing Hazards, Risk and Vulnerability*, Springer Geography, DOI 10.1007/978-94-007-5875-9_5,

depth maps were constructed from satellite and GIS data. Sect. 5.2 describes the evaluation of flood hazards using flood-affected frequency. DEM and HWL data of major rivers in the study area were used to derive multi-temporal flood depth maps (see Chap. 4), which were used to identify hazard areas for the flood depth category (see Sect. 5.3). Finally, in Sect. 5.4, the modeling of flood hazards was conducted by considering a combination of flood depth and flood-affected frequency.

5.2 Analysis of Flood Frequency

Several hydrological parameters, such as the depth of flooding, rate of water-level rises, flood frequency, water velocity, physical exposure of land, and sediment loads, are influencing flood hazards of a particular site. As all parameters were not available, two hydraulic components (e.g., flood frequency and floodwater depth) were considered in this study for the evaluation and determination of potential flood hazard areas.

The concept of flood-affected frequency (Islam and Sado 2000) is used to assess flood hazards. The classified images obtained from the threshold algorithm (see Chap. 4) were used for this purpose. Nine images were classified into water and non-water areas and were lumped together to generate a flood-affected frequency map. Initially, the classified images of 1988, 1998, and 2000 were combined to construct a single flood frequency map. Likewise, the flood extent maps of 2001, 2003, and 2004 were combined, and the process continues for the rest of the years. This procedure resulted in a three-flood extent map, which provided an opportunity to obtain a common boundary essential to developing a flood-affected frequency map (see Fig. 5.1).

The inundation area that appeared in all of the images considered the highly affected areas and therefore a potentially high-damage zone. The common inundated areas that appeared in two and one of the three maps were deemed medium- and low-damage areas, respectively. An inundated area that was not present in any of the images was classified as a non-flooded area (see Fig. 5.2a). The final flood-affected frequency map was reclassified into four categories according to the severity of inundation, corresponding to flood rankings of class 1, class 2, class 3, and class 4 as non-flood, low, medium, and highly hazardous areas, respectively (see Table 5.1).

The flood-affected frequency map revealed that at least 23% of the study area was in a high-hazard zone that is subject to regular inundation. More than 42% of the area was in a non-flooded category that was not inundated by river waters, 13.3% was in a medium-flooded zone, and 20.5% was in a low-flooded category (see Table 5.1).

5.3 Analysis of Floodwater Depth

As described in Chap. 4, a series of flood depth maps were developed using a DEM with the peak water-level data of major rivers. These flood depths were reclassified into four categories: no flooding, shallow, medium, and deep. A rule-based approach was used in the model-builder utility of ArcGIS (v. 10) to derive a floodwater depth

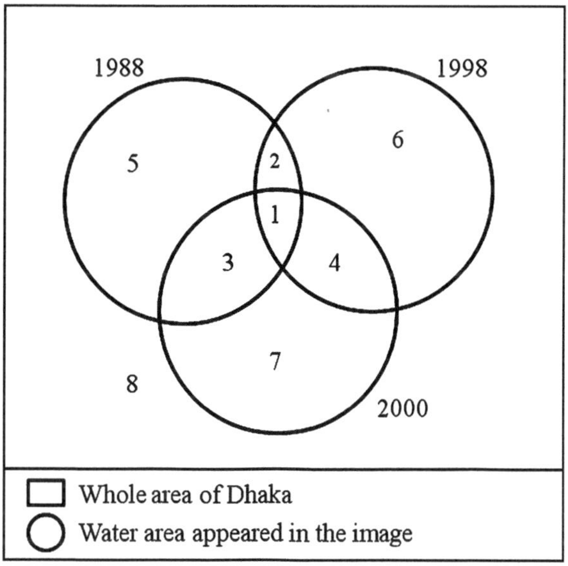

Fig. 5.1 Concept of flood-affected frequency (Adapted from Islam and Sado 2000)

map for the study area. If a deep flood is observed in a single image and a medium flood is observed in at least one of the other images, then it was regarded as deep; from the remaining pixels, if it appeared medium in two images, then it was considered medium. If it appeared shallow in two images, then it was considered shallow. After the classification of the deep, medium, and shallow depth categories, the remaining inundated areas were considered non-inundated or non-flooded. Thus, four floodwater depth categories were obtained from the model, corresponding to depth rankings of class 1, class 2, class 3, and class 4 as non-flooded areas, shallow, medium, and deep (see Fig. 5.2b), respectively.

The total area and percentage occupied by each depth category is shown in Table 5.2. It revealed that 38% of the study area is within the category of no flooding. Shallow, medium, and deep flood depth categories represent 15, 1.5, and 45%, respectively. It should be noted that during the calculation of the depth map, ongoing land development by means of earth/sand filling was not considered. As land development by sand filling is a popular practice that became widespread recently to raise land levels from surrounding river levels, the phenomenon may affect the depth of flooding in the study area. Therefore, the depth maps developed here only represent the condition of land during the creation of the elevation model of the study area by the concerned authority. To validate this study's result, the final flood depth map was superimposed on the flood-affected frequency map, and the difference was found to be only 2% in terms of flood-affected areas. This may suggest the efficacy of these two hydrologic parameters for flood hazard assessment.

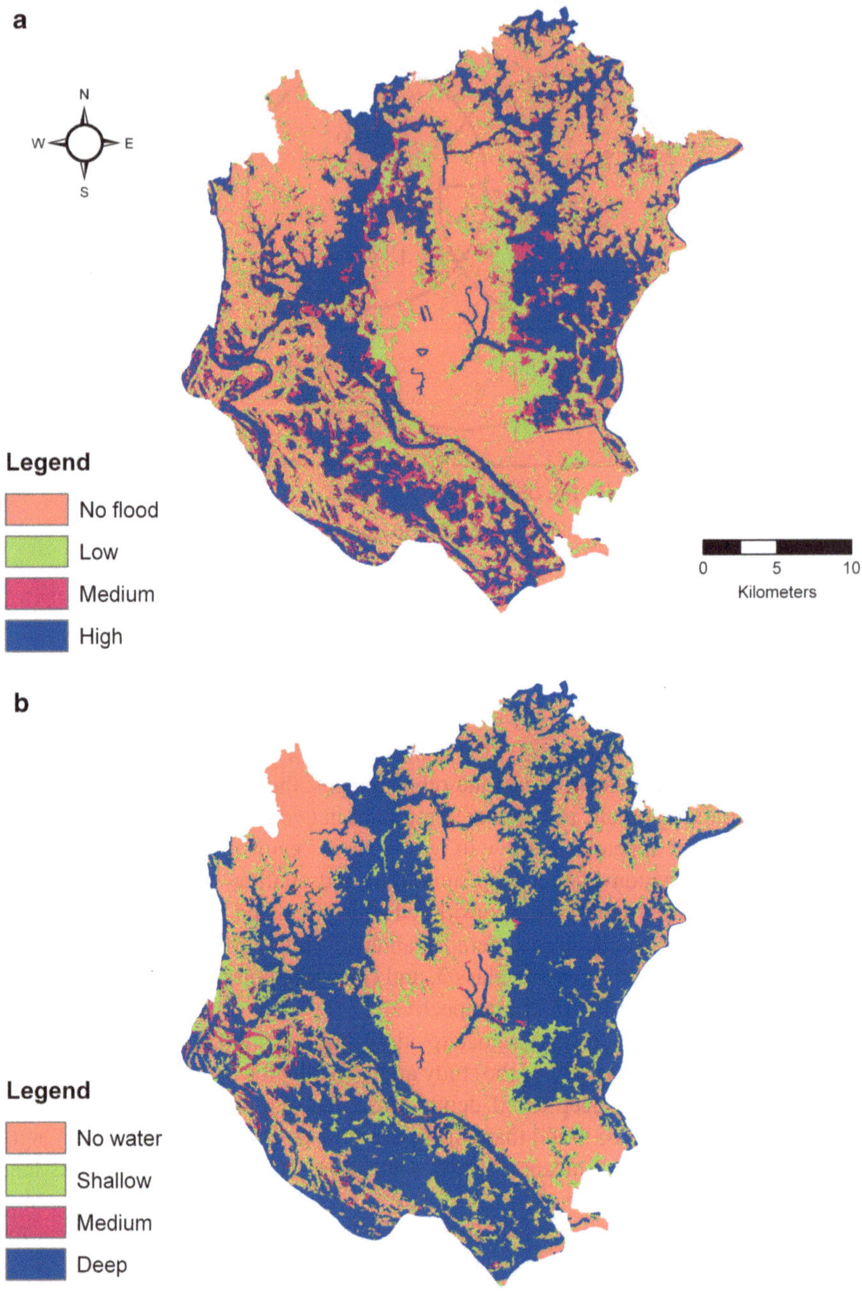

Fig. 5.2 (**a**) Flood-affected frequency and (**b**) floodwater depth

Table 5.1 Classification of flood-affected frequency

SL #	Map1	Map2	Map3	Area (ha)	Area (%)	Class assignment
1	W	W	W	20,680.8	23.5	4
2	W	W	NW	6,684.9	7.6	3
3	NW	W	W	2,746.5	3.1	3
4	W	NW	W	2,253.2	2.6	3
5	W	NW	NW	11,775.1	13.4	2
6	NW	W	NW	5,061.2	5.7	2
7	NW	NW	W	1,229.6	1.4	2
8	NW	NW	NW	37,615.7	42.7	1

Map1 comprises classified maps of 1988, 1998, and 2000; Map2 represents 2001, 2003, and 2004; Map 3 consists of 2005, 2007, and 2009
W water, NW non-water

Table 5.2 Area occupied by each category of floodwater depth

Flood depth category	Area (ha)	Percentage (%)
No flooding	34,053.8	38.5
Shallow	13,100.9	15.0
Medium	1,311.4	1.5
Deep	39,301.5	45.0

5.4 Development of the Flood Hazard Map

Even though geospatial techniques, particularly RS, has shown great potential for hazard assessments (McKean et al. 1991), the big challenge is the lack of generally accepted methods for producing hazard maps (Rhoads 1986). Moreover, many well-developed methods in developed countries may have little relevance to developing countries such as Bangladesh due to a lack of data and/or restricted access to data sources. Therefore, the availability of information and the context of the area being studied are important considerations when evaluating flood hazards. In this study, a simple 2-D multiplication-ranking matrix was applied by combining flood frequency and flood depth maps (Fig. 5.3).

The flood-affected frequency map and floodwater depth maps were combined to derive the final flood hazard map. A 2-D multiplication-ranking matrix (see Table 5.3) suggested by Ochi et al. (1991) was used for flood hazard ranking. The guiding principle of this technique was that flood hazards for a particular cell increase in a nonlinear manner, which depends on both flood occurrence and depth. If a cell represents non-flooded in both maps, then it was considered a non-hazard zone. Low-flooded areas in the flood frequency map and shallow depth in the floodwater depth map were ranked as low-hazard areas. Similarly, medium-flooded areas and medium depth were categorized as a medium-hazard zone. Pixels representing high-flooded areas and deep flooding were considered a high-hazard zone. This operation on two

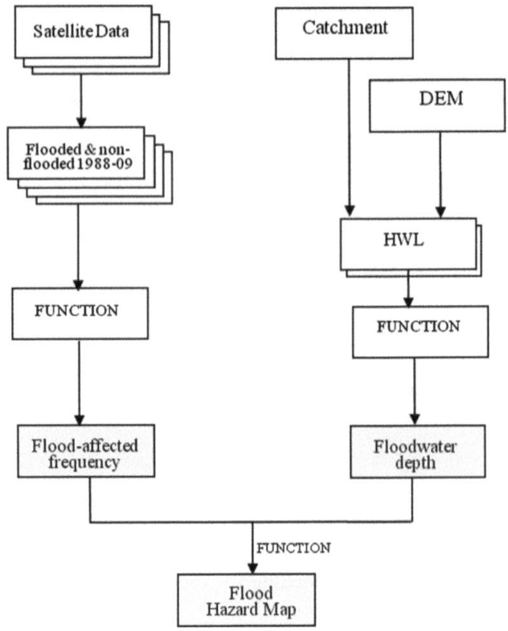

Fig. 5.3 Calculation of flood hazard

Table 5.3 Ranking matrix for two thematic hydraulic parameters

		FD HR			
		1	2	3	4
FAF HR	1	1	2	3	4
	2	2	4	6	8
	3	3	6	9	12
	4	4	8	12	16

FAF flood-affected frequency, *FD* flood depth, *HR* hazard rank

thematic layers resulted in a new map with 16 unique categories. Based on hazard intensity, the derived flood hazard map was reclassified into four hazard categories corresponding to class 1, class 2, class 3, and class 4 as no-hazard, low-hazard, medium-hazard, and high hazard zones, respectively (see Fig. 5.4).

The final flood hazard map was intersected with population and housing data to estimate the exposure of these two elements to flood hazards. Table 5.4 shows the distribution of the population in different flood hazard zones. The analysis revealed that 23.7% of the population was exposed to low–high flood hazards, while 76.3% were located in the no-hazard zone. Among the total population exposed, 24% were elderly, 25% were children, and 24% were females (see Table 5.4). Note that the data used to estimate population exposure are based on the 2001 population census; hence, the actual populations in the different flood hazard zones may be greater than this study.

Exposure of housing in different hazard zones is presented in Table 5.5. This shows that more than 25% of total houses in the study area are exposed to flood

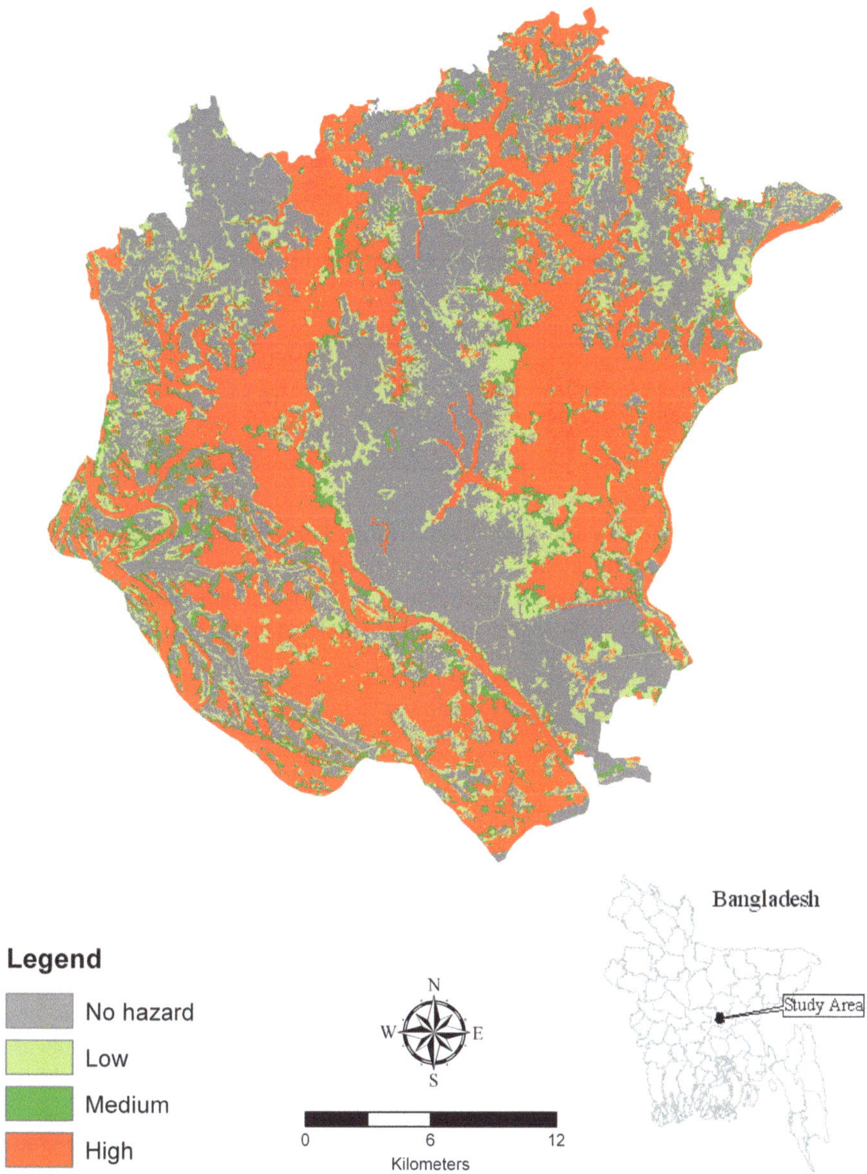

Fig. 5.4 Flood hazard map derived from flood-affected frequency and floodwater depth simultaneously

hazards, of which 6% is exposed to a high-hazard zone. The estimation of various housing types reveals that 17% of *katcha* houses are exposed to medium- to high-hazard categories, while 18% is in the low-hazard zone, revealing the vulnerability of human settlements to floods in Dhaka. Likewise, more than 7% of *semi-pucca*

Table 5.4 Exposure of the population to flood hazards

Hazard zone	Population (%)	Elderly (%)	Children (%)	Female (%)
No hazard	76.3	75.8	74.9	76.0
Low	14.3	14.0	14.9	14.5
Medium	3.5	4.0	3.9	3.6
High	5.8	6.3	6.3	5.9

Table 5.5 Percentage of human settlements in different flood hazard zones

Hazard zone	Total housing units	*Katcha* houses	*Semi-pucca* houses	*Pucca* houses
No hazard	74.1	64.9	79.4	82.1
Low	15.0	18.0	13.4	12.3
Medium	4.8	7.8	3.1	2.3
High	6.0	9.2	4.1	3.3

houses are exposed to medium–high flood hazards. *Pucca* houses, which are usually homes to wealthy people, are also exposed to flood hazards, as shown in Table 5.5. As housing data are based on the year 2004–2005, the actual exposure of human settlements to floods may be higher given that Dhaka is expanding rapidly.

5.5 Summary

This chapter presented the evaluation of flood hazards in the study area by considering flood frequency and flood depth data obtained from GIS and RS images. The study used a simple and cost-effective method of utilizing the GIS to create the flood hazard map from the available dataset.

Using the flood extent maps of 1988–2009, a flood-affected frequency map was constructed, which showed that at least 23% of the area was in a high-flooded zone. In contrast, the flood depth map revealed that 45% of the area was in a deeply flooded zone. Using a 2-D multiplication-ranking matrix, a flood hazard map was derived by considering both flood depth and flood-affected frequency simultaneously.

The exposure analysis revealed that 23% of the total population in the study area was exposed to low- to high-hazard zones. Among the populations exposed, the elderly population constituted 24%, children represented 25%, and females comprised of 24%. This clearly indicates that floods remain a significant threat to the people of Dhaka. In addition, 25% of the total housing units comprising *katcha*, *semi-pucca,* and *pucca* were exposed to flood hazards, of which 6% were in the high-hazard zone. Of the housing types, 18% of *katcha* houses were exposed to the low-hazard zone, 7% were exposed to the medium-hazard zone, and 9% were exposed to the high-hazard zone. As these houses are home to low-income people, the potential effects of floods remain a significant threat for this group. Similarly, 4% of *semi-pucca* houses and 3% of *pucca* houses were exposed to the high-hazard zone, revealing that people in the study area are either not aware of flood hazards or, if they are, they have few alternatives to settling in the high-hazard zone.

As observed in Dhaka, increasing population pressure is forcing many people to move to the vacant land by filling in surface water bodies and floodplains. Consequently, the flood hazard is increasing. To ameliorate flood-induced damage, the derived flood hazard map is invaluable. Urban planners can use this information to make environmentally sound land-use decisions. An important advantage is that the developed hazard map can quickly be updated if modifications occur subsequent to the original study (e.g., changes in land use).

References

Islam MM, Sado K (2000) Flood hazard assessment in Bangladesh using NOAA AVHRR data with geographical information system. Hydrol Process 14(3):605–620

McKean J, Buechel S, Gaydo L (1991) Remote sensing and landslide hazard assessment. Photogramm Eng Remote Sens 57(9):1185–1193

Ochi S, Rahman NM, Kakiuchi H (1991) A study on flood risk evaluation in Bangladesh using remote sensing and GIS. J Jpn Soc Photogramm Remote Sens 30:34–38 (in Japanese)

Rhoads BL (1986) Flood hazard assessment for land-use planning near desert mountains. Environ Manage 10(1):97–106

White GF, Kates RW, Burton I (2001) Knowing better and losing even more: the use of knowledge in hazards management. Environ Hazard 3(3–4):81–92

Chapter 6
Vulnerability and Risk Assessment

Abstract Using census and spatial databases, this chapter demonstrates the assessment of flood vulnerability and risk zoning in the megacity of Dhaka. A place-based approach was adopted to evaluate flood vulnerability at the community level. A method comprising the AHP and the WLC was used to operationalize the conceptual model within a GIS framework. A number of biophysical variables were used to derive the physical vulnerability index (PVI). In addition, nine variables were extracted from diverse sources to derive the social vulnerability index (SVI). To determine the coping capacity of a community, five variables were employed and a coping capacity index (CCI) was developed. Using the PVI, SVI, and CCI, a composite vulnerability index (CVI) was prepared, which revealed that 28 and 14% of the population in the study area were located between high and very high vulnerable zones, respectively. Thirty-six percent were in *katcha* housing units, which are largely home to marginal people and were in the very high vulnerable zone, revealing the susceptibility of human settlements to floods. Flood hazard data were multiplied with the CVI to delineate the flood risk zone. A total of 18.5% of people lived in high and very high flood risk zones, which was further compounded by the number of females, young, elderly, and illiterate persons. At least 25.8% of housing units were located in these zones. The flood risk map developed in this study could be used as baseline information for the development of pertinent mitigation measures to ameliorate flood loss in the days ahead.

6.1 Introduction

Assessing vulnerability to natural hazards such as flood is a critical starting point to minimize disaster-related damage to society. As megacities are the locus of a large concentration of population and infrastructures, the likelihood of damage from natural hazards is also high. Many studies are now focusing on understanding megacities' vulnerability to natural hazards, which encompasses the systematic identification

A.M. Dewan, *Floods in a Megacity: Geospatial Techniques in Assessing Hazards, Risk and Vulnerability*, Springer Geography, DOI 10.1007/978-94-007-5875-9_6,

of vulnerable communities and properties at risk (World Bank 2011). Although multiple definitions of vulnerability exist in literature, a general consensus is that the vulnerability of a particular community depends on a number of factors, including physical, social, political, and economic (see Chap. 2). Risk can have different meaning in different disciplines; however, the quantification of flood risk enables emergency managers and local governments to prepare against an imminent disaster (Takeuchi 2001).

Community-based disaster risk management has received considerable attention since the late 1980s because of the problems associated with top–down approaches (Bollin and Hidajat 2006; UNISDR 2004; UN 2003; Maskrey 1989). Risk maps based on neighborhoods and groups are important for the effective management of emergencies and crises (Morrow 1999), particularly for identifying the risk characteristics of a community (Birkmann 2007), capturing the root causes of vulnerability (Fekete et al. 2010; Azar and Rain 2007), and making informed decisions about risk reduction (Flax et al. 2002). Although geospatial techniques have shown immense potential in mapping the risk and vulnerability attributes of a community (Tran et al. 2009), relatively few researchers have integrated physical, social, and economic variables within a GIS to explore the potential of this technique in developing countries (Wang et al. 2011). Based on existing hazard-related literature, this study conceptualizes a framework for understanding community vulnerability and associated risk from floods in the megacity of Dhaka (see Chap. 2). Using a range of variables at a fine scale, this empirical study attempts to characterize vulnerability and the risk of floods. As a number of procedures for flood risk assessments exist, this study employs spatial multi-criteria methods that include analytic hierarchy process (AHP) and weighted linear combination (WLC) techniques to operationalize the concept developed (see Chap. 2).

This chapter presents the methods and results of vulnerability and risk estimation. Section 6.2 presents the data, and detailed methods of vulnerability estimation are presented in Sect. 6.3. The variables were derived from both geospatial data and census geography, and the vulnerability to flood was evaluated in two distinct areas—physical and social—and they are presented in Sects. 6.3.1 and 6.3.2. The capacity of a community to withstand or recover from floods is analyzed and explained in Sect. 6.3.3. The derivation of overall vulnerability is elucidated in Sect. 6.4, and risk assessment is illustrated in Sect. 6.5. Finally, the results and discussions are presented in Sect. 6.6.

6.2 Analytical Techniques

6.2.1 Data

Four variables were used to assess the (bio)physical vulnerability, as the locational factor is an important indicator to depict the susceptibility of humans and properties to floods. A reasonable assumption is that people living in the proximity of active channels or in properties on a floodplain are at a higher risk than those living in highly elevated areas or away from rivers. Therefore, physical variables have been used to estimate (bio)physical vulnerability. Data regarding physical vulnerability assessments were collected from a number of sources. A DEM was collected from the SOB (see Chap. 4), geology and river network vector files were collected from CEGIS, and land-use/cover data were obtained from literature (Dewan and Corner 2012).

To derive the indicators for social vulnerability and coping capacity, a range of variables was constructed from diverse sources. Socioeconomic and demographic attributes were collected from the BBS community series (BBS 2003), which represent the population census of 2001. These data were encoded in a spreadsheet and then linked with the census tract boundary (see Chap. 4) using a unique ID. A number of variables, such as population density and sex ratio, were subsequently calculated for each community. Types of housing and road network data were collected from the DAP of RAJUK (see Chap. 4). Land value data between 2008 and 2012 were collected from the respective sub-registry offices, and only the average residential land prices were used in this study. It should be noted that some of the census tracts do not have land value data for 2012. In this situation, 2008 residential land value data were used. Poverty data were obtained from the Bangladesh Household Income and Expenditure Survey 2005 (BBS et al. 2005). Other variables, such as the location of hospitals and flood shelters, were derived from the DAP and SPARRSO databases. In addition, multi-temporal flood maps classified from RS data (see Chap. 4) were used to compute the community flood awareness variable. As the spatial databases that were developed in this study derive from a variety of sources, a method suggested by Lo and Faber (1997) was used for the integration of census data with other databases, and a Bangladesh Transverse Mercator (BTM) system was used as the referencing system.

6.3 Vulnerability Assessment

A variety of techniques are available to compute community vulnerability to natural hazards, and they can be categorized as inductive or deductive (Yoon 2012). Similarly, variables that are used to derive the vulnerability of an area depend on various factors such as scale of analysis, extent of the study area, and the availability of data (Fekete et al. 2010). For instance, indicators that are usually applied to compute vulnerability in developed countries are largely lacking in developing nations; hence, the choice and accessibility of data remain a grave concern. Nevertheless, a vulnerability map derived from different indicators is typically represented by an index to depict the degree of susceptibility of a community to natural hazards. As mentioned above, this study considers indicators from physical, socioeconomic, and other sources to examine community vulnerability to floods. Note that the physical and social indicators were analyzed separately to produce an individual index and the results were then combined to construct a composite vulnerability index

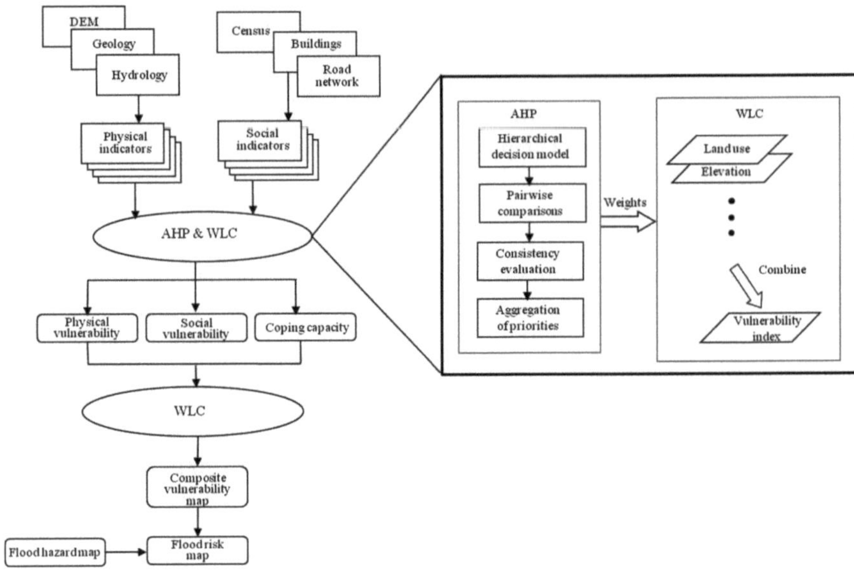

Fig. 6.1 Flowchart showing overall methods of the study

(CVI). The basic premise of the work was to create flood vulnerability and risk maps that can easily be interpreted and to allow improved decision making for mitigating flood-related damage in Dhaka.

A place-based approach was adopted to estimate the vulnerability of communities by integrating spatial and census data. Spatial multi-criteria assessment techniques in the form of AHP and WLC were used as a method of analysis (see Fig. 6.1). The AHP was first introduced by Thomas L. Saaty as a robust and flexible technique for supporting priority setting and improved decision making when the phenomenon that is being studied involves both the quantitative and qualitative aspects of a decision (Saaty 1977). Although this technique can be used to check the consistency of preferences (DeSteiguer et al. 2003), flood risk assessments based on these techniques are rare (Meyer et al. 2009) (see Chap. 2). As disaster risk reduction initiatives display a special challenge at the megacity level (Wisner 2003), consideration of AHP and WLC techniques would allow some of the associated problems to be overcome. The following sections detail the procedures of the adopted techniques.

The AHP is based on multi-criteria that prioritize the criteria identified by different groups of people (stakeholders and experts) involved in the decision-making process in order to arrive at the best decision. It also assists in justifying the optimality of the decision (Saaty 1980). The AHP allows a problem to be organized as primary and secondary objectives, which is known as the hierarchy, and a matrix is subsequently used to weigh each factor against every other factor within each level of the hierarchy. Every level in the process is tied with its top and bottom levels, resulting in a clear priority statement for a group or individual (Ramanathan 2001). The following steps are involved in the process:

(a) Structuring a hierarchical decision model
(b) Development of a pairwise comparison matrix
(c) Obtaining local priorities and checking consistency of comparison
(d) Aggregation of local priorities

The hierarchical decision model is the design phase of AHP in which the top level exhibits the overall goal of the decision. Indicators can be criteria in the upper level of the model, and each of them is further broken down into sub-criteria. Each pair of criteria or sub-criteria element is compared in terms of its relative importance using a 9-point system from 1 (if two indicators equally contribute to the objective) to 9 (when one indicator is strongly favored over another to meet the objective) (see Table 6.1) and forms a comparison matrix, namely, a pairwise comparison (see Table 6.2). A score of 1 denotes equal importance, a score of 3 refers to weak preference, while scores 5 and 7 represent obvious and strong preferences. The even numbers (i.e., 2, 4, 6, and 8) are used when a compromise is needed between the odd numbers.

The local priority (weight) for a criterion is computed from a pairwise comparison matrix by normalizing the points in the columns (divide a cell value by the sum of a column) and averaging the normalized points in the row of the criterion. The consistency of the comparisons is evaluated by calculating a consistency ratio (CR). If the CR is equal to or less than 0.1, the comparisons are considered consistent, otherwise it would be revised. The CR is defined by the following equation:

$$\text{CR} = \text{Consistency Index} / \text{Random Index} \tag{6.1}$$

Table 6.1 Semantic scale of the AHP method

Comparative importance	Definition	Description
1	Equal importance	Two indicators equally influence the parent decision
3	Weak importance	One factor is moderately influential over the other
5	Essential or strong importance	One factor is strongly favored over the other
7	Demonstrated importance	One decision factor has significant influence over another
9	Absolute importance	Evidence favoring one decision factor over the other is the highest order of affirmation
2, 4, 6, 8	Intermediate	When compromise is needed, values between two adjacent judgments are used
Reciprocals	If A_i is the judgment value when i is compared with j, then A_j has the reciprocal value when compared to A_i	A reasonable assumption

Source: Ramanathan (2001)

Table 6.2 Example of a pairwise comparison matrix

	Population density	Age	Gender
Population density	1	3	5
Age	1/3	1	4
Gender	1/5	1/4	1

Table 6.3 Random average consistency indexes for various *n*

n	1	2	3	4	5	6	7	8	9
RI	0.0	0.0	0.58	0.90	1.12	1.24	1.32	1.41	1.45

n represents the number of criteria

The random index (RI) refers to a randomly generated reciprocal matrix from the 9-point scale, and it can be obtained by referring to the RI table. Saaty (1980) provides a function of *n* in relationships (see Table 6.3). The consistency index (CI) is defined as

$$CI = (\lambda_{max} - n)/(n-1) \tag{6.2}$$

where λ_{max} is the largest eigenvalue derived from the comparison matrix and *n* is the number of criteria. Once the consistency is validated, the final priority of the criterion at the upper level of the hierarchy model is obtained by aggregating the local priorities of the criteria at its lower level.

In addition to AHP, WLC is also used in this study. Due to its simplicity, multi-criteria decision analysis in terms of WLC is a frequently used technique within a GIS (Malczewski 2006, 2004, 1999, 1996). In GIS analysis, each indicator is treated as a data layer. A WLC is conducted by multiplying indicators by the corresponding weights and aggregating all weighted layers. In this study, the following equation was used:

$$VI = \sum_{i=1}^{i} \sum_{j=1}^{j} W_i W_{ij} x \tag{6.3}$$

where VI is the vulnerability index for physical or social indicators, w_i and w_{ij} are the weights for the *i*th and *j*th sub-criterion of the *i*th criterion, and *x* represents the value of an indicator.

6.3.1 Analysis of Physical Vulnerability

This study asserts that flood risk in Dhaka is not only governed by socioeconomic factors but is also significantly influenced by physical/natural factors. Therefore, four biophysical variables were considered to derive a PVI (see Fig. 6.2). The distance to active channels was calculated by using river network data, assuming that people who live close to active channels have an elevated risk compared to those who do not. The Euclidian distance function was used to derive the distance to the active channel. Elevation categories were derived by slicing the DEM data. Geology,

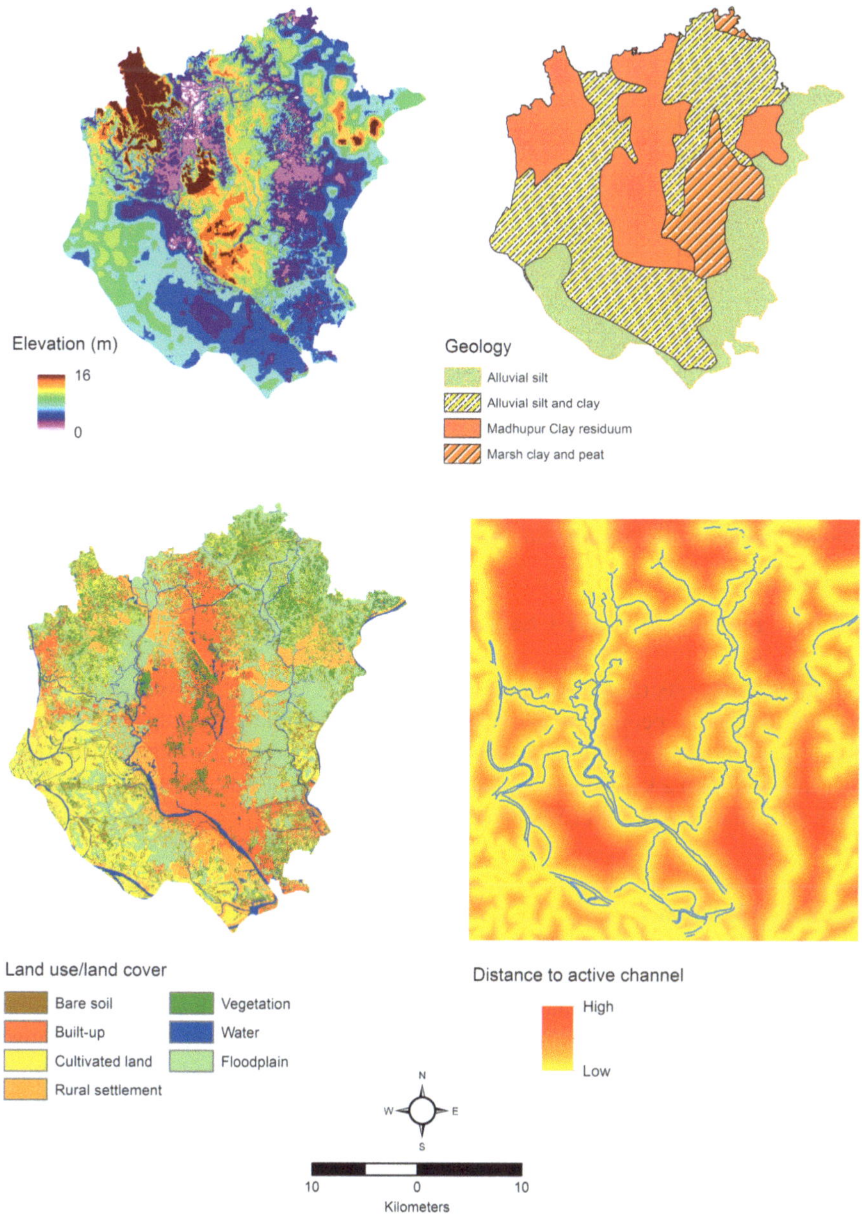

Fig. 6.2 Biophysical variables

land use, elevation, and distance to active channels were considered for estimating physical vulnerability. For the sub-criteria of distance to the active channel and elevation category, the intensity of importance was determined based on the relationships between flood hazard categories and elevation and distance to active channel parameters.

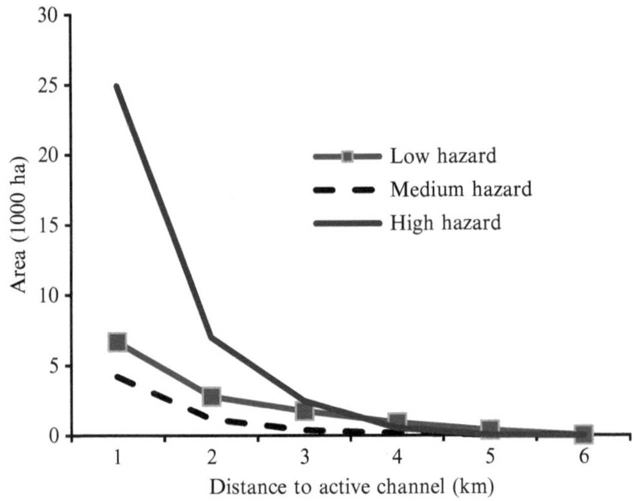

Fig. 6.3 Distance categories according to different flood hazard classes, by area

6.3.1.1 Distance to Active Channel

Figure 6.3 demonstrates that areas within 2–4 km of river networks are highly hazardous; that is, people and property are in considerable danger up to 4 km away from the existing river networks. Hence, the maximum weight was allocated to the distance less than 2 km from rivers, and the lowest weight was given to the category with a distance greater than 5 km (see Table 6.4).

6.3.1.2 Elevation

Elevation is the most important criteria because it has a tremendous effect on flood hazard in the study area, as most areas in Dhaka are low-lying lands. The elevation category was intersected with the flood hazard map to ensure that the relative weights for each category were justified properly. As expected, the high-hazard category is between the elevation of 1 and 4 m (see Fig. 6.4). Therefore, elevation of less than 4 m was given the highest weight, while elevation greater than 9 m was assigned to the lowest weight (see Table 6.4).

6.3.1.3 Land Use/Cover

The land-use/cover data comprises of seven categories: built-up, bare soil or landfill, cultivated land, rural settlements, vegetation, wetlands, and water bodies. Since the potential effects of floods are high for the built-up category in terms of infrastructural

Table 6.4 Biophysical vulnerability decision hierarchy model

	2		3	
1	Criteria	Weight	Criteria	Weight
Physical vulnerability index	Land-use category	0.316	Built-up	0.484
			Rural settlements	0.273
			Cultivated lands	0.144
			Vegetation and bare soil	0.051
			Wetlands and water bodies	0.048
	Distance to active channel (km)	0.145	<2	0.432
			2–3	0.295
			3–4	0.165
			4–5	0.070
			>5	0.038
	Elevation (m)	0.483	<4	0.482
			4–5.5	0.299
			5.5–7	0.122
			7–9	0.064
			>9	0.033
	Geology	0.056	Alluvial silt and clay	0.474
			Marshy land	0.134
			Alluvial silt	0.320
			Madhupur terrace	0.072

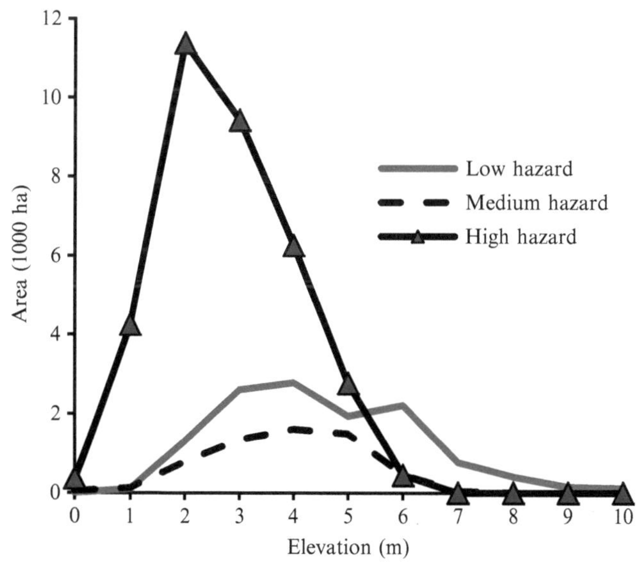

Fig. 6.4 Elevation in each hazard zone, by area

damage and human sufferings, relative weights were allocated highest for this land-use class. Considering the effect on humans, rural settlements were assigned to the second highest weight. Subsequent weights were allocated to cultivated land, as agricultural loss due to floods has a considerable impact on the local economy. Wetlands and water bodies were allocated the lowest weight because they do not pose a threat to people; instead, they act as water retention ponds during floods (see Table 6.4).

6.3.1.4 Geology

Geologically, the area can be divided into four categories. The Madhupur terrace is a highly elevated area, and it is not usually inundated during the wet season. Therefore, it was assigned to the lowest weight. Alluvial silt and clay are areas where flood is a normal phenomenon. This was given the highest importance, followed by the alluvial silt category (see Table 6.4). In addition, marshy land is normally flooded during monsoons, and this type of geological category is not inhabited. Hence, it received lower importance in the weightings.

Regarding the importance of the criteria at the upper level (level 2), elevation was considered the most important criteria and given the highest weight. The second most important was given to land use because of rapid urbanization, and consequent land-use change is exacerbating river water flooding in the study area. Moreover, the potential effects of flood are high because built-up areas comprise human settlements and urban infrastructures. Flood hazard distribution reveals that the central and northern portions are not usually inundated, as land elevation is relatively high and is protected with embankments and raised roads. Therefore, two other parameters (distance from rivers and geology) are given less importance in the second decision level. The pairwise comparison matrices with CR to evaluate the consistency of the comparison judgments can be found in Appendix I.

After deriving the weights for each variable using the AHP, WLC was used so all variables could be weighted and combined to derive a PVI. Before performing the WLC, the values for elevation and distance to active channels were standardized using the below equation:

$$p = \frac{\max - x}{\max - \min} \tag{6.4}$$

where p is the standardized value, min and max represent the minimum and maximum values for each dataset, respectively, and x is the cell value.

Finally, WLC was performed first on the sub-criteria of each indicator and subsequently on the criteria at the upper level of the decision hierarchy model. The resulting index indicated that the higher the value, the greater the susceptibility to flood of a community.

6.3.2 *Analysis of Social Vulnerability*

An efficient social vulnerability assessment for environmental hazards requires good baseline data when the spatial unit includes the local level of analysis (Cutter et al. 2000, 2003, 2009). As the study is based on the lowest level of census geography, nine variables were extracted from the census and relevant spatial databases to determine the human dimensions involved. These variables were divided into three categories: demographic, socioeconomic, and infrastructures and lifelines (see Figs. 6.5, 6.6, and 6.7). It should be noted that the poverty data are available at the *thana* level. As the spatial unit of study is at the community scale, poverty data were subject to areal interpolation. Using the method suggested by Holt et al. (2004), poverty data were interpolated into the community level. The variables were first used to assess their importance in understanding vulnerability using the AHP technique (see Fig. 6.8). Note that the strength of importance in the pairwise comparisons for the criteria was determined on the basis of literature.

6.3.2.1 Population Density

Floods affect human populations in many ways, for example, physically and psychologically. Therefore, population density is regarded as one of the most important indicators in determining social vulnerability. Crowdedness caused by high density may introduce many problems during and after a flood event, including evacuation difficulties (Chakraborty et al. 2005; Cova and Church 1997) and the increased risk of disease transmission (Hewitt 1997; Anderson 1992). The parameter is also used to assess "elements at risk" while estimating flood risk (Tingsanchali and Karim 2005). Therefore, the population density indicator was given higher importance at all levels of the decision hierarchy. It was calculated as the total number of people residing in a census tract divided by its total area (km^2). Higher importance was allocated for higher density, while lower importance was given to lower density (see Table 6.5).

6.3.2.2 Children and Elderly Population

Studies have shown that children and the elderly are highly vulnerable to natural hazards (see Wisner et al. 2004) because of their restricted mobility and difficulty with evacuation during emergencies (Paul 2010; Ngo 2001; Kaniasty and Norris 1999; Chowdhury et al. 1993; O'Brien and Mileti 1992). A total number of populations for these groups was extracted and classified into four categories: aged 0–4, 5–9, 10–14, and over 60. As shown in Table 6.5, children in the 0–4 age group in the study area were given maximum weight in the lowest decision hierarchy because this group constitutes around 35% of the city's population (Hossain 2008). The 5–9 and 60+ age groups were given subsequent importance in the pairwise comparison (see Appendix II).

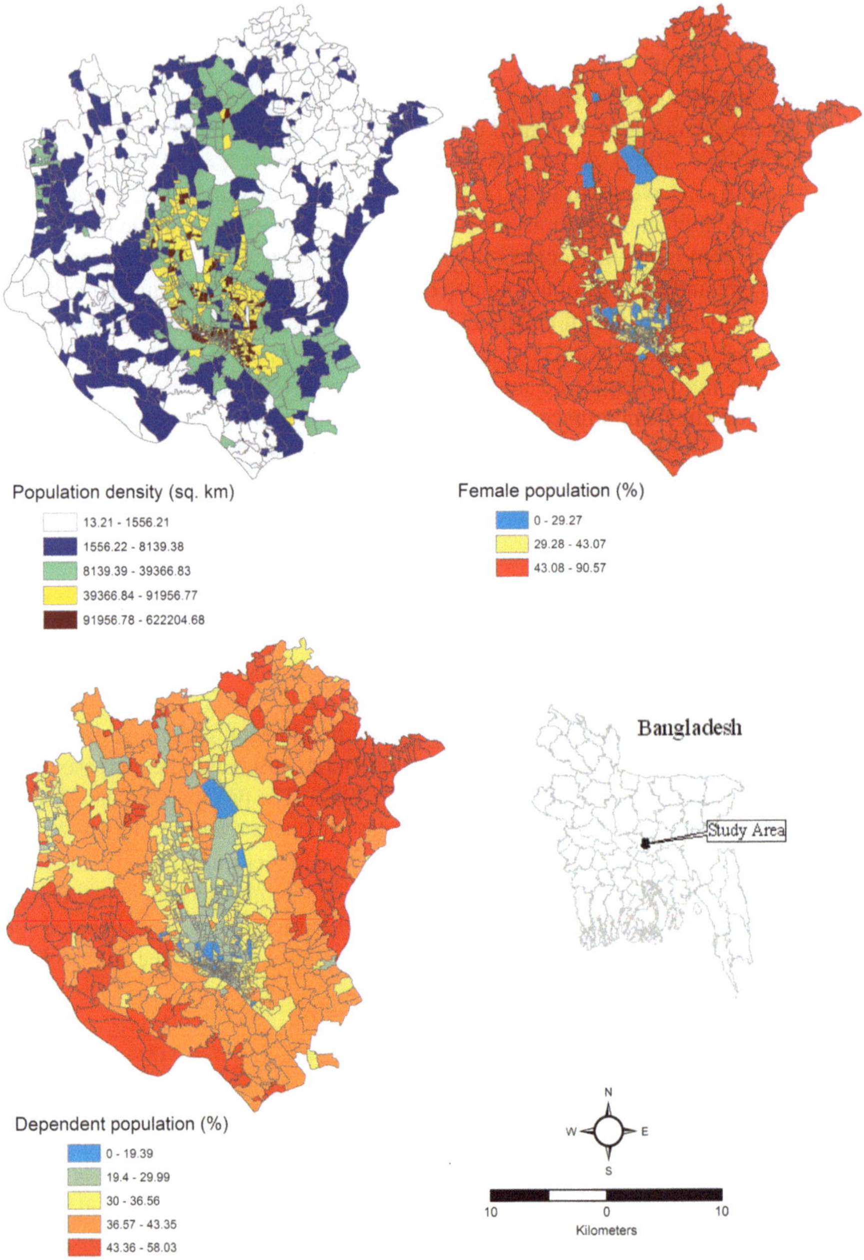

Fig. 6.5 Demographic indicators

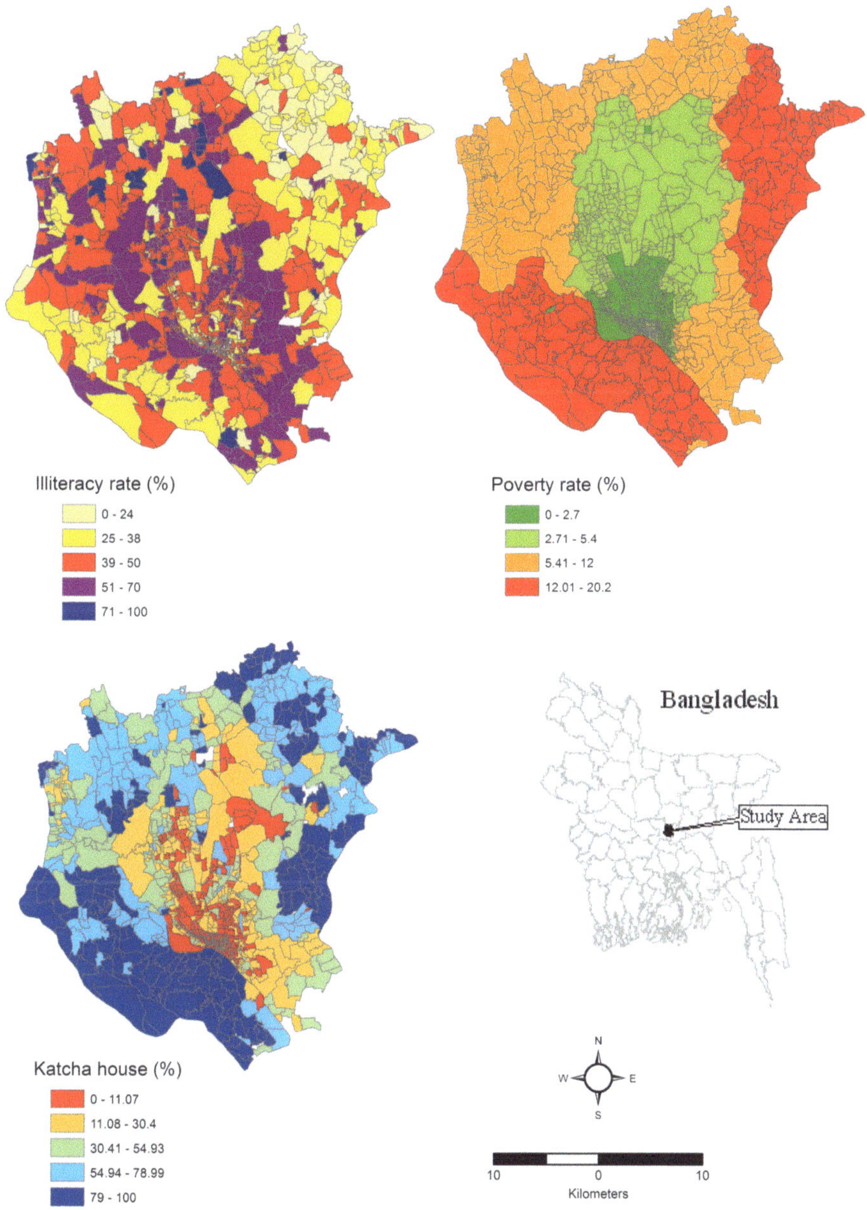

Fig. 6.6 Distribution of socioeconomic indicators

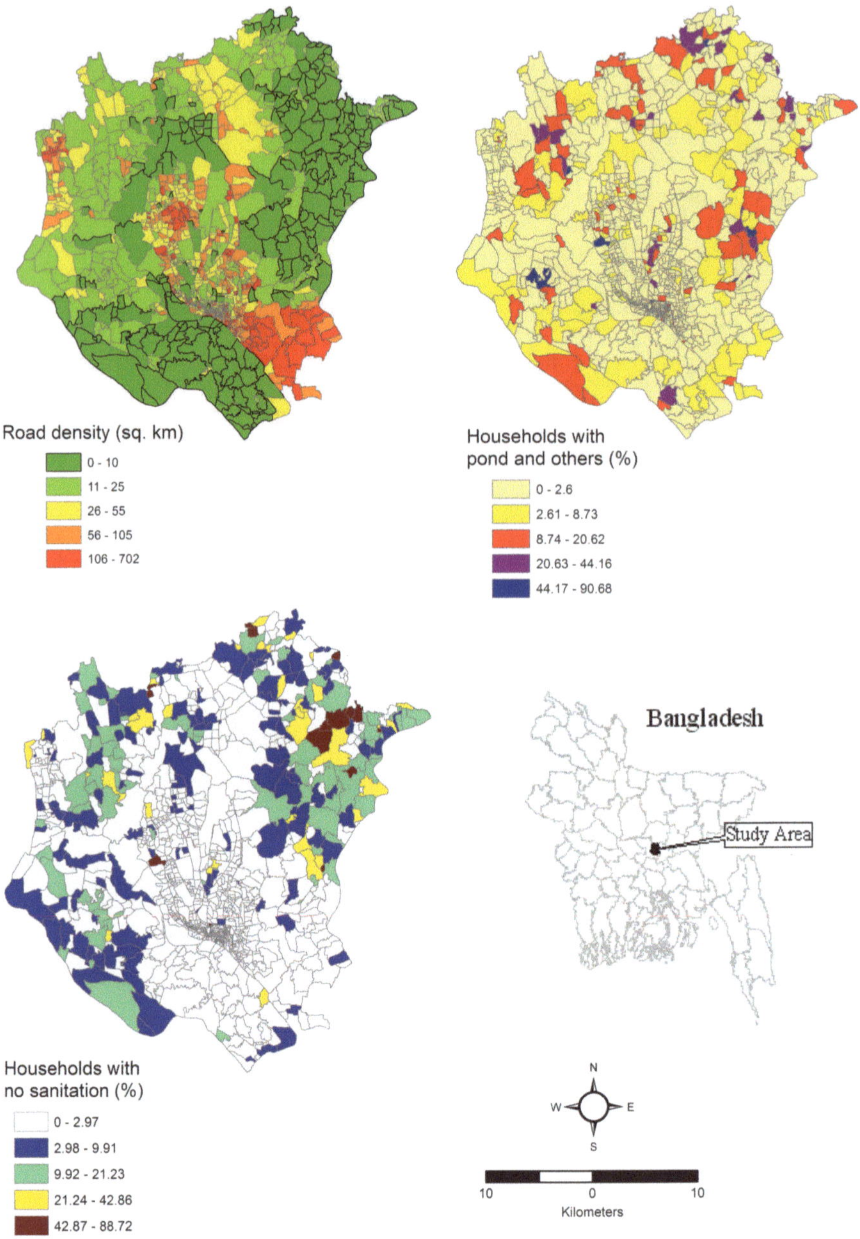

Fig. 6.7 Distribution of lifeline and infrastructure indicators

6.3.2.3 Gender

Gender is an important consideration when depicting vulnerability to natural hazards (Cutter et al. 2003; Liverman 1990), particularly in developing countries where sociocultural and religious restrictions inhibit the mobility of women (Sultana

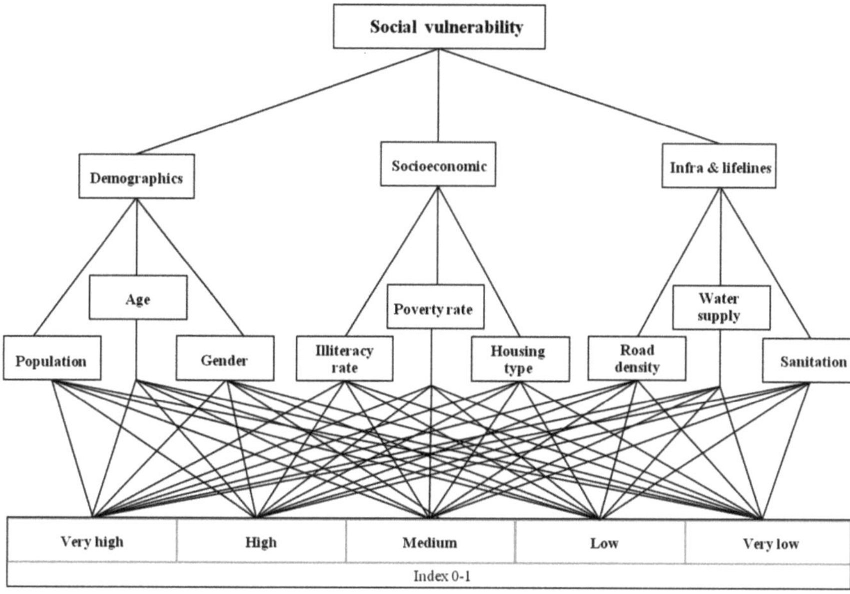

Fig. 6.8 Decision hierarchy model for social vulnerability assessment

2010; Ray-Bennett 2009; Sultana and Thompson 2008; Khatun 2005; Hutton and Haque 2004; Khondker 1996; Cannon 2002; Begum 1993; Agarwal 1990). Moreover, most females in Bangladesh devote their lives to household activities from an early age and have fewer opportunities to participate in educational and social activities than males (Islam and Sultana 2006), making the female population highly vulnerable to natural hazards. As a result, hazard-related mortality is also higher among women than men (Ikeda 1995; Chowdhury et al. 1993; Begum 1993). Therefore, higher importance was given to females in the gender variable (see Table 6.5).

6.3.2.4 Illiterate Population

This variable is particularly important when assessing vulnerability, as educated people are generally more aware of extreme events and they play an important role in the reduction of damage caused by flood (Paul and Routray 2010). The number of illiterate males and females was used with the assumption that illiterate people are dependent on others to prepare and evacuate themselves during an impending disaster. Moreover, an illiterate person does not have the ability to read or write, which is a significant impediment during emergencies. Thus, the census districts with the highest illiteracy rates were given the highest importance when allocating weights in the decision hierarchy (see Table 6.5).

Table 6.5 Decision hierarchy model for socioeconomic and infrastructure vulnerability

	2		3		4	
1	Criteria	Weight	Criteria	Weight	Criteria	Weight
Social vulnerability index	Demographic	0.557	Population density (km^2)	0.619	<30,000	0.034
					30,000–85,000	0.067
					85,000–165,000	0.133
					165,000–350,000	0.245
					>350,000	0.521
			Age (year)	0.284	0–4	0.597
					5–9	0.231
					10–14	0.047
					>60	0.125
			Gender	0.097	Male	0.250
					Female	0.750
	Economic	0.320	Illiteracy rate	0.143	<24	0.036
					24–38	0.069
					38–50	0.136
					50–70	0.262
					>70	0.497
			Types of housing (%)	0.301	*Katcha*	0.690
					Semi-pucca	0.244
					Pucca	0.066
			Poverty (%)	0.484	<2.7	0.053
					2.7–5.4	0.112
					5.4–12.0	0.253
					>12.0	0.582
	Infrastructure and lifelines	0.123	Road density (km^2)	0.126	<50	0.042
					50–100	0.075
					100–200	0.134
					200–400	0.264
					>400	0.485
			Water sources	0.458	Tap	0.041
					Tube well	0.079
					Well	0.118
					Pond	0.362
					Others	0.400
			Sanitation	0.416	Sanitized	0.078
					No sanitation	0.688
					Others	0.234

6.3.2.5 Housing Types

The quality of housing is a critical indicator when assessing vulnerability for a given area (Satterthwaite 2010; Kubal et al. 2009; Pelling 1997; ESCAP 1991). As the quality of houses is substandard and the density of buildings is extremely high, damage caused by natural hazards is likely to be enormous in Bangladesh, particularly in Dhaka (Paul and Bhuiyan 2010; Ansary 2003). Moreover, houses built on floodplains and in close proximity to water bodies tend to exhibit greater vulnerability. As mentioned in Chap. 4, three types of housing information are available in the study area: *katcha*, *semi-pucca*, and *pucca*. These housing types were incorporated into the vulnerability evaluation. As *katcha* houses are poor in quality and subject to severe damage from floods, they were given higher importance than the other two categories in the pairwise comparison matrix (see Appendix II).

6.3.2.6 Poverty

The linkage between hazards, vulnerability, and poverty are complex (Few 2003); however, many studies assert that low-income people suffer the most from natural hazards (Dasgupta 2007; Burton et al. 1993; Davis 1987). For example, an account of the total number of deaths from floods between 1975 and 2000 revealed that the low-income group had the highest number of deaths, while high-income groups accounted for around 1% (UN 2003). The study further corroborated that 50% of the deaths occurred in the low-income category, while 49% occurred in the middle-income group. This clearly shows the association between poverty and flood vulnerability. As shown in Chap. 1, rampant poverty in urban areas in developing countries poses a serious threat to the effective management of natural hazards; therefore, a poverty indicator should be incorporated into the vulnerability assessment. Bangladesh's poverty data comprises both the higher and lower rates of poverty at the *thana* level; however, this study uses the lower poverty rate. As shown in Table 6.5, higher importance and subsequent weighting were allocated to the greater proportion of poor people in each community, and the lowest weight was given to the lowest proportion.

6.3.2.7 Road Density

Infrastructure facilities such as roads, air, and rail networks are in the category of "lifeline" (Platt 1995 in Cutter et al. 2000) and play a significant role in evacuation (Rasid et al. 2000) and post-event relief and recovery (Sanyal and Lu 2006, 2009). As these networks are important in determining the lifeline of a community, the density of roads was calculated by intersecting the road network feature dataset with the community boundary feature and weighted using the AHP (see Table 6.5).

6.3.2.8 Water Sources

Access to safe drinking water is a key element for flood risk management. Several studies demonstrated that the outbreak of water-borne disease is widespread during floods due to the increased transmission of pathogens (Tapsell et al. 2002; Kondo et al. 2002). As noted in Chap. 3, surface water pollution has already exceeded standard limits in the study area due to increasing anthropogenic activities.

Therefore, water sources in the study area become severely contaminated in the wet season (Quraishi et al. 2005; Khan et al. 2005; Sohel et al. 2003; Rahman et al. 2002; Islam and Yusuf 2001), posing serious health risks to the inhabitants of Dhaka (Dewan et al. 2012). There are a number of water sources for drinking purposes, namely, taps, tube wells, wells, ponds, and rivers, and pure drinking water becomes scarce following the floods. Therefore, other water sources received the highest weighting, as people relying on this source are likely to be at a greater risk of water-borne disease. Pond and well categories were allocated with subsequent importance, as they are likely to be contaminated earlier than tap and tube well waters (see Table 6.5).

6.3.2.9 Sanitation

Similar to other parts of the country, the sanitation of Dhaka is not encouraging, as millions of people are at risk of communicable diseases after a flood event (Rahman et al. 2002, 2004; Rahman and Bux 1995; Siddique et al. 1989). Due to the recurring nature of floods, sanitation data per households from the census was taken into account and classified as access to safe sanitation, no sanitation, and other means of sanitation. The no sanitation category was given the most weight, followed by other means of sanitation (see Table 6.5).

Appendix II presents the pairwise comparison matrices for demographics, socio-economic, and infrastructure and lifeline criteria with their corresponding consistency ratios. The census values for the indicators were first converted into ratios from count, and subsequently the weights were assigned to the ratio. The normalization was then performed using the below equation:

$$s = \frac{x - \min}{\max - \min} \tag{6.5}$$

where s is the normalized value, min and max represent the minimum and maximum values of the indicator variable, respectively, and x is the weighted value of the community. This operation resulted in an index between 0 and 1. The derived social vulnerability index (SVI) indicated that the higher the value, the greater the vulnerability of a particular community.

6.3.3 *Analysis of Coping Capacity*

To develop the interventions required for effective disaster management, it is important to analyze the status quo of a community in terms of its capability to endure the effects of environmental hazards (Anderson and Woodrow 1998). This would allow emergency managers to systematically identify the communities with the least capacity to sustain during floods and subsequently prioritize the need to improve their capability to cope with extremes. A range of qualitative techniques (e.g., questionnaire survey) is traditionally utilized to examine people's capacity to cope with natural hazards. This is one of the areas in which indigenous adjustments and coping mechanisms have been extensively examined in Bangladesh (Braun and Aßheuer 2011; Paul and Routray 2010; Rayhan 2008; Brouwer et al. 2007; Khandker 2007; Dewan 2006; Younus et al. 2005; Choudhury et al. 2004; Ali 2004; del Ninno et al. 2003; Paul 1984, 1995, 2003; Rashid 2000; Hasan et al. 1999; Dewan et al. 1999; Baquee 1997; Rahman 1996; Dodson 1996; Rasid and Mallik 1995; Hasan et al. 1999; Zaman 1993; Alam and Chowdhury 1992; Islam 1990; Alam 1990; Shaw 1989; Alam and Samsuddin 1988; Das 1988; Islam 1974). However, these techniques suffer from a number of shortcomings. They require an enormous amount of time and skilled personnel. The study is often confined to a particular segment of a community, thereby disregarding a certain spectrum of the population, which is essential to understand vulnerability of the entire community to natural hazards. For example, people located in remote areas are generally omitted from the study design due to constraints such as inaccessibility. In addition, perception and opinion research, or so-called *victim-centered* studies, may produce skewed results, which could be a vital issue for effective flood management (Cook 2010). These shortcomings may be subdued by using a more systematic method, including the use of census and spatial data that enables the identification of the coping capacity of a particular community over space, thereby permitting more informed decision making.

The selected indicators for coping capacity analysis are the total literacy rate, average housing price, per capita hospitals, per capita flood shelters, and flood awareness (see Fig. 6.9), as shown in the decision hierarchy model (see Fig. 6.10). For these indicators, higher importance was given to the sub-criteria, with higher values in the pairwise comparison computation so that areas with better coping capacity could be mapped.

6.3.3.1 Mean Housing Value

As neither household income nor median housing prices are available at the community level, mean housing values were estimated using mean residential land prices and the types of housing as a surrogate variable. All three types of residential houses—*katcha*, *pucca*, and *semi-pucca*—were taken into account. Three assumptions were made to carry out this analysis: (1) people residing in *katcha* housing are poorer than the other two categories and more vulnerable to floods; (2) residents of *pucca* housing are affluent and able to cope with natural extremes; and (3) middle and lower-middle classes have access to *semi-pucca* housing, and the degree of vulnerability is lower than the marginal group.

Figure 6.11 presents the estimation of mean housing prices. The residential building feature dataset was first intersected with the community polygonal database. For each of the intersected buildings, the value was estimated based on its dimensions and the corresponding land value per decimal (40.46 m^2). Weights were allocated based on the types of houses and the three assumptions explained above. The highest

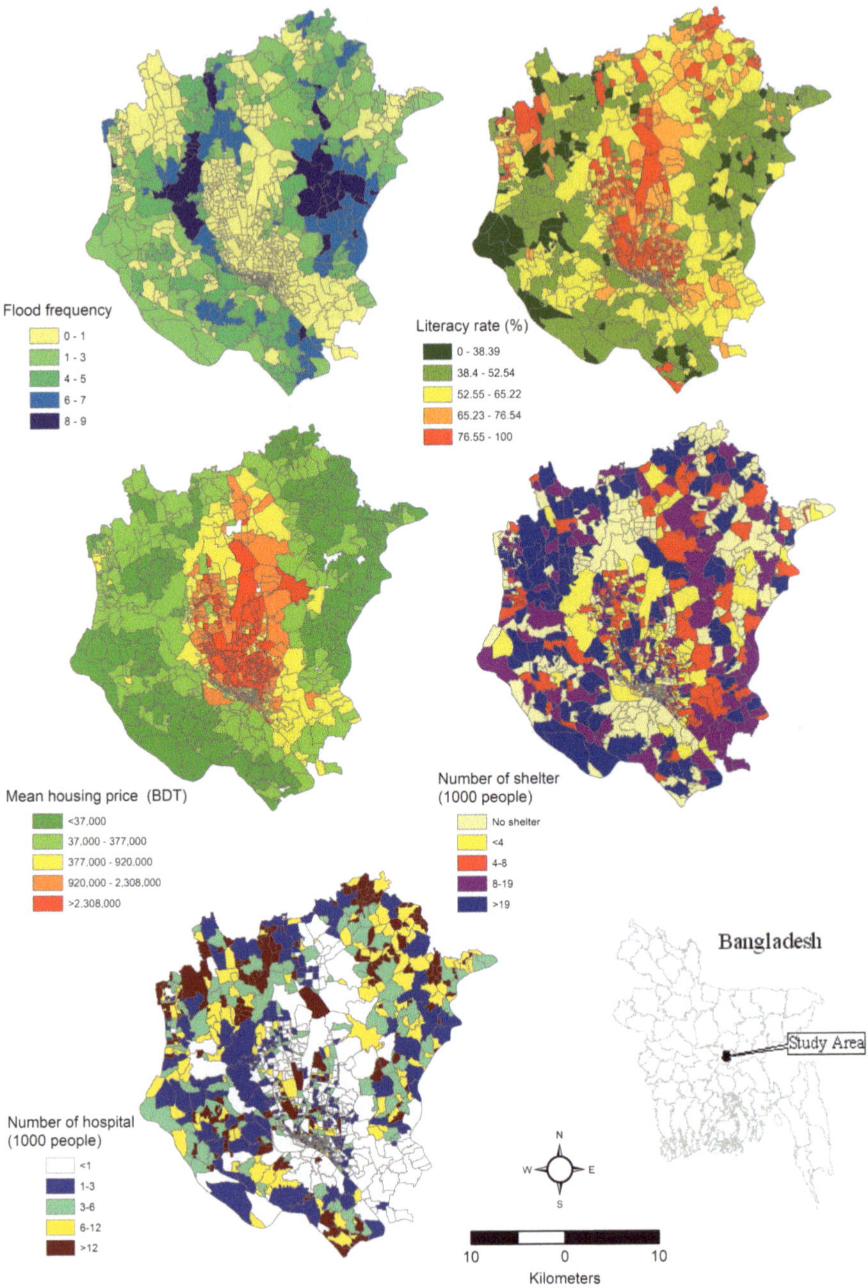

Fig. 6.9 Distribution of coping capacity indicators

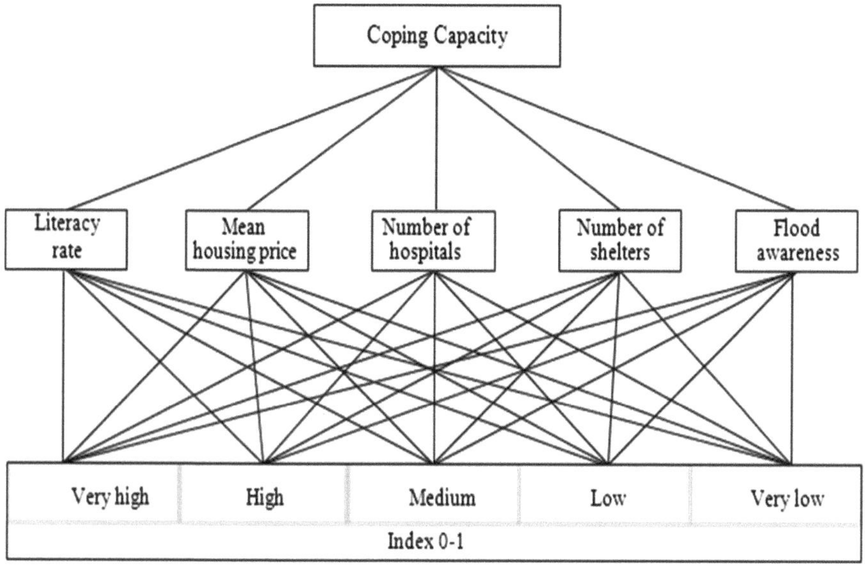

Fig. 6.10 Decision hierarchy model for coping capacity analysis

weight was assigned to the value of *pucca* housing, and the lowest value was given to *katcha* housing because of its construction materials. The average housing value was estimated for each community. Note that the resulting value does not indicate the actual real estate or market price of a typical residential house in Dhaka; however, it can be used as a proxy in the absence of actual median/mean housing price data.

6.3.3.2 Literacy Rate

Education is an important human wealth that allows people to make appropriate decisions and take pertinent mitigation measures to withstand or recover from natural disasters (D'Oyley et al. 1994). It also affects people's lifetime earnings (Cutter et al. 2003). Literature suggest that households with higher levels of education cope better than households with lower levels of education (Paul and Routray 2011). To examine this proposition, the total literacy rate was used and weighted for each community to assist in understanding their coping capacity (see Table 6.6). The highest weight was assigned to communities with the highest literacy rate for both males and females, and the lowest weight was assigned to those with lower literacy rates.

6.3.3.3 Per Capita Shelters and Hospitals

The availability of shelters and health-care facilities in hazard-prone areas is a vital indicator of coping capacity of a particular community, a factor that gained

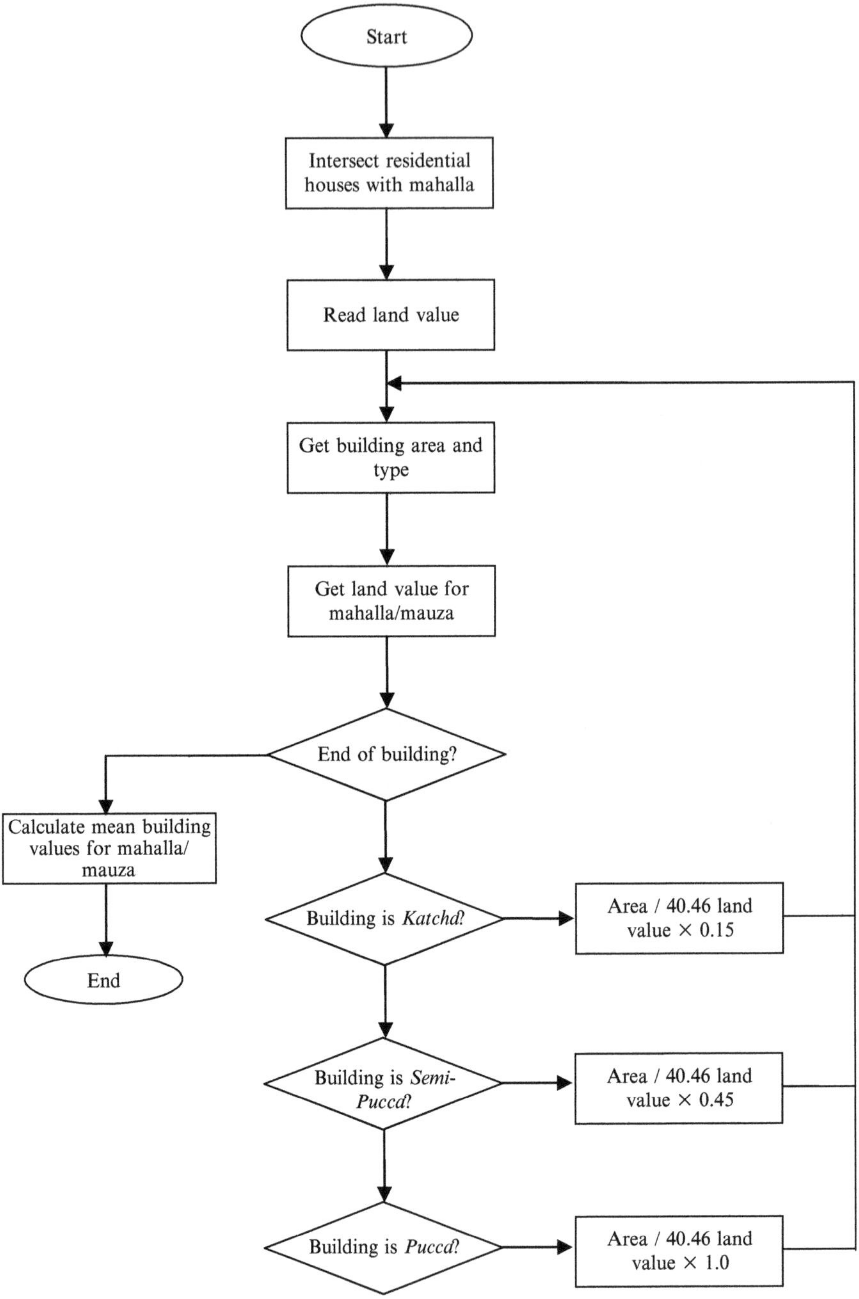

Fig. 6.11 Estimation of mean housing price

Table 6.6 Decision hierarchy model for coping capacity criteria

	2		3	
1	Criteria	Weight	Criteria	Weight
Coping capacity	Literacy rate (%)	0.267	<38.0	0.040
			38.0–52.5	0.074
			52.5–65.0	0.142
			65.0–76.5	0.248
			>76.5	0.496
	Mean housing price ('000 BDT)	0.160	<26.8	0.034
			26.8–200.09	0.063
			200.09–1,017.3	0.120
			1,017.3–3,737	0.273
			>3,737	0.510
	Number of hospitals (per 1,000 persons)	0.040	<3	0.038
			3–10	0.054
			10–30	0.137
			30–100	0.280
			>100	0.491
	Number of shelters (per 1,000 persons)	0.069	<2	0.039
			2–5.7	0.056
			5.7–13	0.134
			13–25	0.269
			>25	0.502
	Flood frequency (awareness)	0.464	<3	0.062
			3–5	0.108
			5–7	0.267
			>7	0.563

significant importance for flood management and disaster preparedness planning (Sanyal and Lu 2009; Khan 1991). Unlike cyclone-prone areas on the coast of Bangladesh, shelters in the study area usually comprise of community centers, schools, religious, and other education centers (Alam and Ali 2002; Maniruzzaman and Alam 2002). A database containing the locations of educational institutes, hospitals, and community and religious centers was prepared using the DAP and SPARRSO databases to calculate per capita shelters and hospitals in the study area. The higher the number of shelters and hospitals in each community, the higher the weight in terms of the third level of the decision hierarchy; however, the highest weight was given to the per capita shelter indicator in the second level of the decision hierarchy (see Table 6.6).

6.3.3.4 Flood Awareness

The classic natural hazard paradigm shows that prior experience of hazards plays a crucial role in mitigating the negative consequences of floods and preparing against disasters (Paul 1997). Therefore, the flood frequency variable was derived by overlaying multi-temporal flood maps from 1988 to 2009 (see Chap. 4) on the community database, with an assumption that people experiencing floods every year are more prepared for future events than communities with little knowledge of flood (Botzen et al. 2009). The highest weight was given to this variable among other coping capacity variables in the second level of the decision hierarchy, as disaster preparedness is an important factor for a community to cope with natural disasters. It should be noted that the pairwise comparison for each variable is provided in Appendix III; however, the aggregated weights are shown in Table 6.6.

Equation 6.5 was used for coping capacity variables to standardize before conducting WLC. The resulting CCI (Coping Capacity Index) indicates that the higher the index, the better the coping capacity of a particular community.

6.4 Derivation of Composite Vulnerability Index (CVI)

A composite vulnerability map was produced by combining vulnerability and coping capacity maps using WLC technique. An equal weighting scheme was adopted for this purpose (Cutter et al. 2003). To combine three maps and estimate CVI, first of all, mean PVI score for each community was derived by intersecting it with the community boundary file. Then, the CCI was linearly inversed so that the lower the CCI value, the higher the contribution of a particular community to flood vulnerability. The following equation was used to derive the overall vulnerability or CVI:

$$\mathrm{CVI} = 0.33 * \mathrm{PVI} + 0.33 * \mathrm{SVI} + 0.33 * \mathrm{CCI} \tag{6.6}$$

It should be noted that the inversed $\mathrm{CCI} = \max - \mathrm{CCI}$.

6.5 Flood Risk Assessment

A flood risk map is one of the most important measures for preventing flood loss potential (Yoshino and Yoshikawa 1985). A flood risk map based on fine spatial units (e.g., at the community level) is particularly useful for planners and emergency managers when developing appropriate countermeasures. Flood risk is defined in this study as a function of flood hazard and vulnerability (Apel et al. 2009; Wisner et al. 2004). The following equation is used to derive flood risk map in the study area:

$$\text{Flood risk} = \text{hazard} \times \text{vulnerability} \tag{6.7}$$

To compute flood risk, the mean hazard value for each community was obtained by overlaying the flood hazard map (see Chap. 5) with the community feature dataset. The value was then normalized using Eq. 6.5, which produced a range of 0–1; thus, the hazard map can correspond with the value of the CVI. Finally, the hazard map was multiplied with the composite vulnerability map to derive the spatial distribution of flood risk in the study area. The resulting flood risk map was reclassified into five classes—no risk, low, moderate, high, and very high—according to the intensity of flood risk; thus, the map supports visualization of flood risk in the study area.

6.6 Results and Discussion

6.6.1 Physical Vulnerability

The PVI map revealed that areas closer to water bodies with a low elevation constitute higher vulnerability to flood (see Fig. 6.12). Although central part of the city, including the central business district, is in the low vulnerable zone (0.00–0.33), some areas in the old part of Dhaka are in the medium vulnerable zone (0.51–0.64). Generally, the older part of Dhaka and its adjoining areas are the focal point of economic activities and are densely populated, with a good segment of people are in the low-income group; hence, the vulnerability of humans and property is considerably high. Areas that are projected to be urbanized in the next few years, such as those in the southern and eastern directions, are in the high and very high vulnerable zones (0.63–1.00). Due to their low elevation and close proximity to river networks, communities in these areas are highly susceptible to extreme flood loss (see Table 6.7). For instance, more than 42% of the population is vulnerable with respect to their locational attribute, as mean distance from active channels is very low—between 0.5 and 0.6 km. Their vulnerability is further exacerbated because the mean elevation of the land is also very low (3.6–3.8 m). A small percentage of the population (3.7%) is in the very low physical vulnerability zone. Flood-control structures influence the hydrological regime of a particular area; therefore, some areas appear to have elevated physical vulnerability due to the changes in the hydrological flow caused by embankments and other flood-control works. This supports the earlier results of hydrodynamic modeling (Masood and Takeuchi 2012), which revealed that the locational factor is an important indicator in estimating flood vulnerability in Dhaka.

6.6.2 Social Vulnerability

As described above, a number of indicators were used to determine social vulnerability of the communities. The resulting map revealed that few communities, particularly those with higher housing prices, are in the category of very low social vulnerability (<0.23), implying that people in this zone are able to absorb losses relatively quickly. Interestingly, many communities in the flood protection zones exhibit medium levels of social vulnerability (0.33–0.44). It is worth noting that

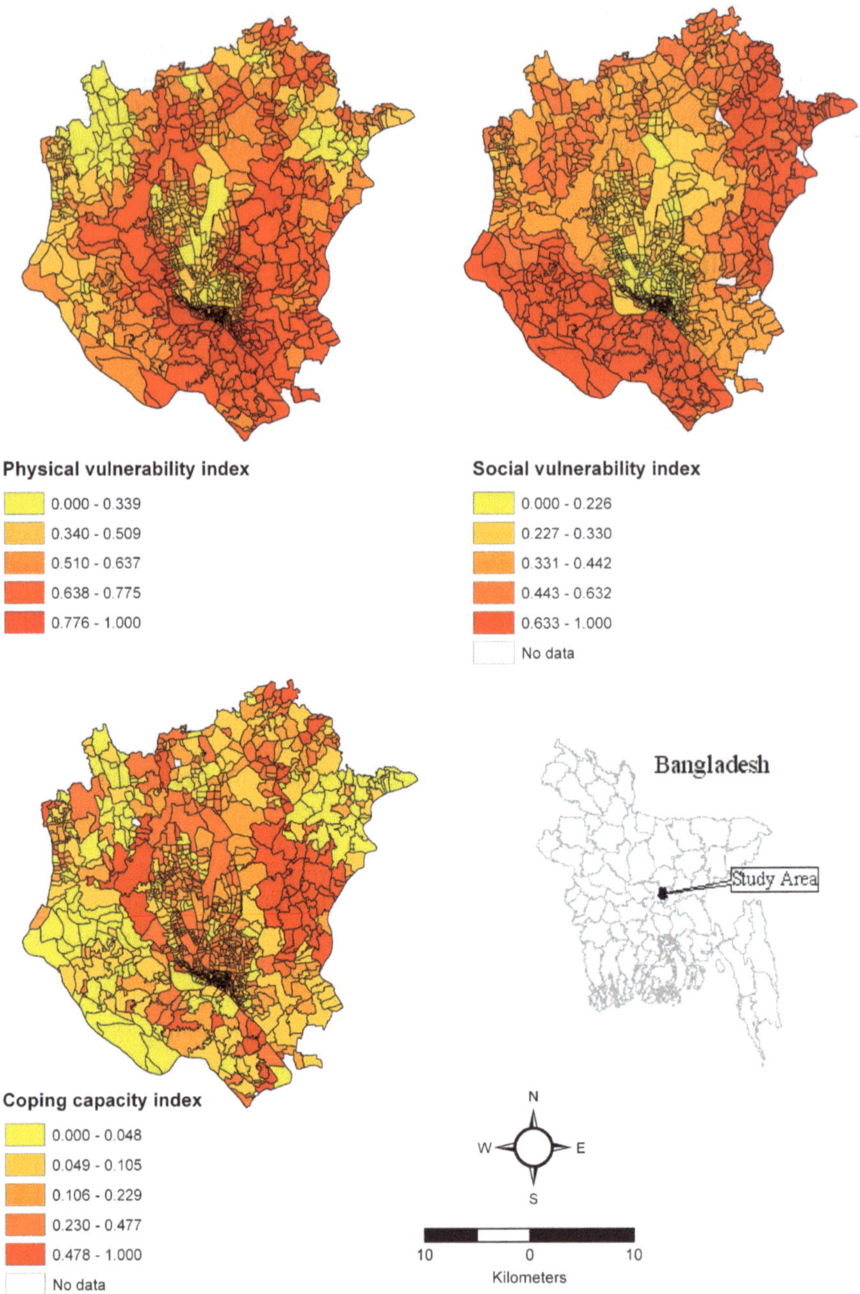

Fig. 6.12 Spatial patterns of physical and social vulnerability, and coping capacity, by community

Table 6.7 Distribution of variables in relation to physical vulnerability

PV zone[a]	Population (%)	Mean elevation (m)	Mean distance to active channel (km)
Very low (<0.34)	3.7	7.5	2.0
Low (0.34–0.51)	26.2	7.1	1.7
Medium (0.51–0.64)	27.2	5.8	0.9
High (0.64–0.77)	23.0	3.6	0.6
Very high (>0.77)	20.0	3.8	0.5

[a]Physical vulnerability

people in protected areas tend to misperceive their risk of flooding, which could increase long-term vulnerability (Etkin 1999) and deepen the risk of flood disaster (Tobin and Montz 1997). There is a higher likelihood of water-borne diseases during and after floods in the medium vulnerability zone, as 33.8% of total households do not have access to safe sanitation, and 33% mainly rely on ponds and other means of water sources.

Very high vulnerable zones (>0.63) are distributed along the fringe areas, and Table 6.8 explains that the social vulnerability of high and very high vulnerable zones is due to high levels of poverty and a large number of *katcha* houses—with the exception of mean population density. Although mean population density is comparatively low in these zones, other variables, such as the rate of poverty, are high, and the percentage of *katcha* houses is extremely high (see Table 6.8). In contrast, access to safe water and sanitation is very low, suggesting a degree of vulnerability to floods. The areas that are currently considered for urban expansion are in the extreme vulnerable zone (see Fig. 6.12), signifying the potential for higher flood losses in the context of intensified precipitation driven by climate change. Moreover, existing flood-control works are based on a stationary climate; therefore, these structural measures will no longer be applicable (Milly et al. 2008) in the context of intensified precipitation. As a result, both pluvial and fluvial floods could turn into catastrophes if appropriate measures, such as restricting development in the floodplains, are not taken into account. The results obtained here could greatly support emergency managers and local governments in their efforts to lessen damage from future floods.

6.6.3 *Coping Capacity*

The derived coping capability map showed different results compared to those of vulnerability maps, partly because of the integration of the flood awareness variable into the derivation of the coping capacity of communities. However, a coping capacity indicator can be conceptualized as a form of the adaptation capacity of individuals or a community in the study area (Few 2003). As the majority of people in the built-up zone did not experience floods during the past few events, they are in

Table 6.8 Distribution of variables regarding the social vulnerability zone

				Water sources			Sanitation	
SV zone[a]	Mean population density (km^2)	Poverty (%)	*Katcha* houses (%)	Tap (%)	Tube well and well (%)	Pond and others (%)	Sanitized (%)	Others and no sanitation (%)
Very low (<0.23)	24,338.8	2.2	0.4	12.8	0.09	2.5	9.4	1.1
Low (0.24–0.33)	24,957.3	3.0	10.0	55.1	18.2	30.6	43.7	24.1
Medium (0.34–0.44)	9,199.4	5.7	29.6	27.7	42.7	33.0	33.8	37.0
High (0.45–0.63)	3,385.5	7.5	18.9	2.2	16.0	21.5	6.4	15.5
Very high (>0.64)	2,654.2	16.7	41.2	2.1	22.2	12.9	7.6	22.4

[a]Social vulnerability

the low (0.05–0.11) to medium (0.11–0.23) coping capacity zone. Communities that experience floods almost every year exhibit a very high coping capacity (>0.48) despite their low per capita hospitals and shelters, which supports the suggestion made by other studies (Paul and Routray 2010, 2011). The central portion of the city and the relatively economically affluent areas also show a high coping capacity (0.23–0.48) because these areas comprise of a higher concentration of shelters, hospitals, and high literacy rates (see Fig. 6.12).

6.6.4 Composite Vulnerability

This study conceived that the coping capacity is part of vulnerability assessment (UNDP 1992) and should be included in the estimation of the overall/total vulnerability of a particular area. A CVI was therefore mapped out by combining PVI, SVI, and CCI (see Fig. 6.13) and further reclassified according to the severity of vulnerability. The map was subsequently intersected with various indicator maps (e.g., population) to discern human dimensions involved. A total of 399 census tracts (communities) were found to be highly vulnerable to floods, while 170 communities are in the low zone of vulnerability (<0.03). The population distribution in each zone indicated that 28 and 14% of people are located in the high and very high zones of vulnerability, respectively. As the literacy rate is lower, this further compounds the degree of vulnerability to flood in both zones. Further, the percent of *katcha* houses, which are usually home of poor people, is higher in the very high vulnerable zone (36%). In contrast, the number of *semi-pucca* houses, which are home of middle- and lower-middle-class households, is highest in the high vulnerable zone (Table 6.9). With the forecast increase in Dhaka's population, the city's flood vulnerability will increase; therefore, it is important to understand the spatial distribution of vulnerable people and property, which could allow an efficient management of future floods.

6.6.5 Flood Risk Mapping

Flood risk map was derived by multiplying the mean hazard score and composite vulnerability map. While hazard represents physical processes, flood vulnerability indicates susceptibility to damage and the risk of human lives.

To understand the human dimensions of flood risk, population and other parameters were intersected with the derived risk map, and statistical analysis was conducted. Although most of the elevated places and built-up areas are between no risk and low risk zones, areas that are likely to be urbanized in the next few years are at the higher end of the risk spectrum (see Fig. 6.14). Using population census data from 2001 and the socioeconomic distribution, Table 6.10 summarizes vulnerable subgroups into different flood risk zones. A total of 18.5% of the population lives

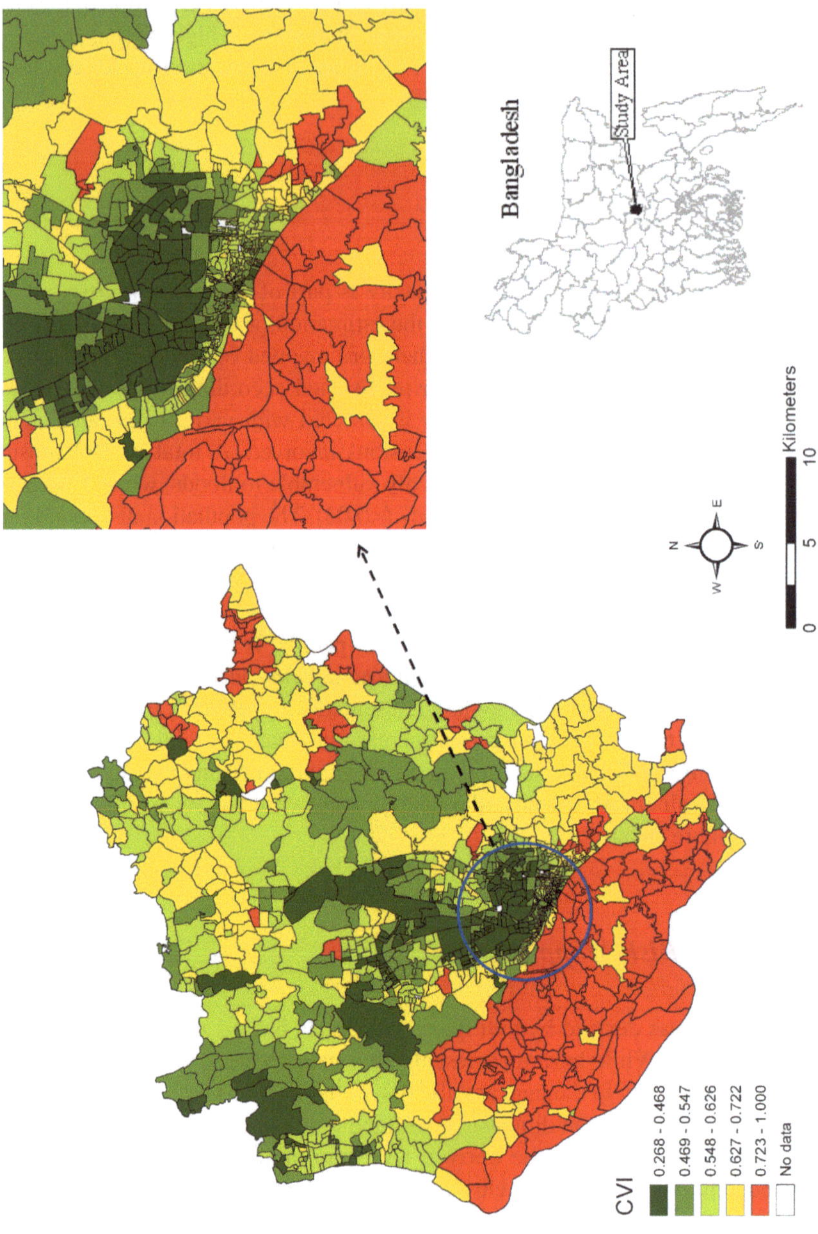

Fig. 6.13 Spatial patterns of the CVI, by community

Table 6.9 Distribution of variables relating to CVI

CVI[a] zone	Population (%)	*Katcha* houses (%)	*Semi-pucca* houses (%)	*Pucca* houses (%)	Mean literacy rate	Mean housing value (BDT)[b]
Very low (<0.47)	12.0	2.8	6.0	13.9	81.6	6,327,255.4
Low (0.47–0.55)	21.0	11.8	16.6	23.7	70.5	2,062,260.6
Medium (0.55–0.63)	24.5	21.9	30.7	27.2	63.0	954,770.3
High (0.63–0.72)	28.2	27.2	37.4	26.1	58.1	752,246.7
Very high (>0.72)	14.3	36.4	9.3	9.2	49.6	212,920.2

[a]Composite vulnerability index
[b]Bangladeshi currency, that is, Taka

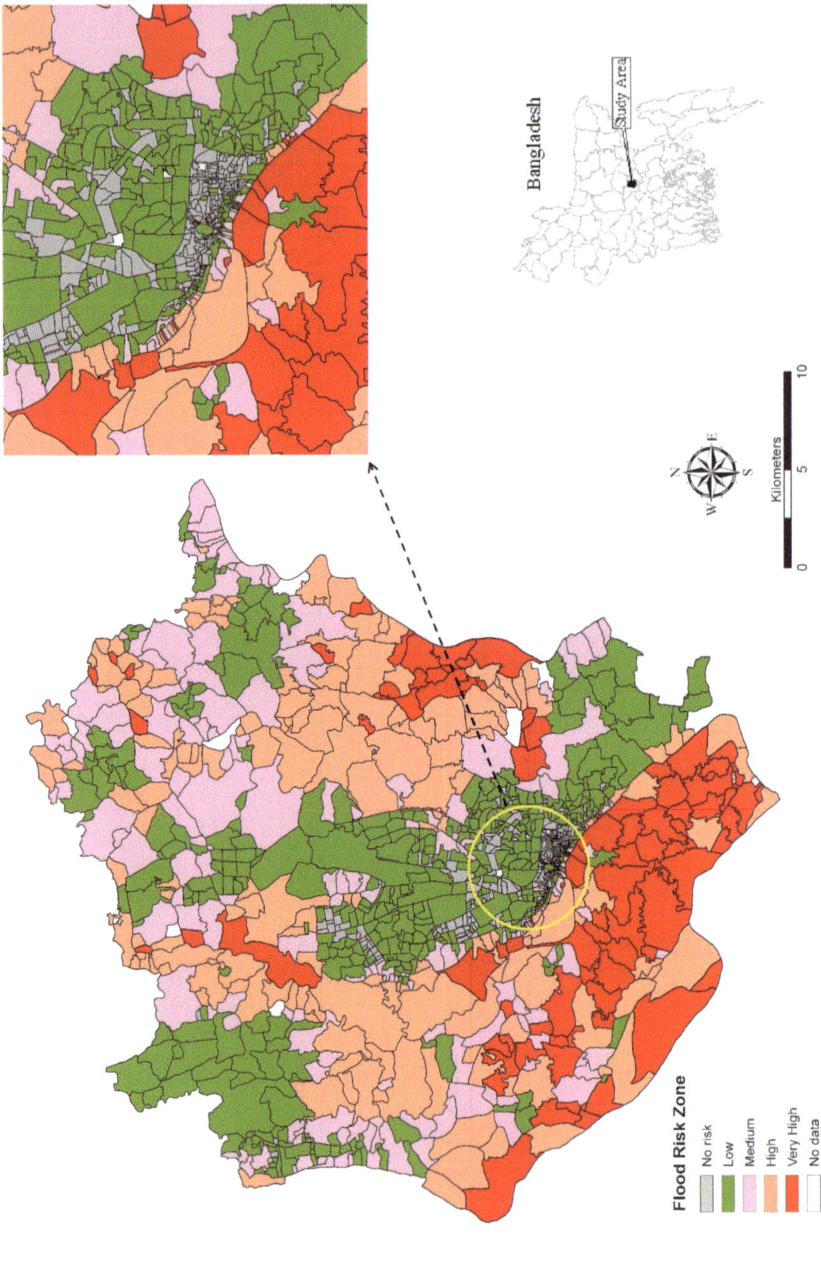

Fig. 6.14 Flood risk map, by community

Table 6.10 Distribution of variables in different flood risk zones

Flood risk zone	Population		Female		Young (0–14 years)		Elderly (>60)		Illiterate persons		Housing units	
	Number	%	Number	%	Number	%	Number	%	Number	%	Number	%
No risk	1,301,590	16.1	555,508	15.5	340,135	13.9	50,343	16.3	108,781	13.6	52,486	7.0
Low	4,049,379	49.9	1,773,278	49.5	1,183,922	48.2	141,524	45.9	384,490	48.0	345,307	46.0
Medium	1,256,022	15.5	562,435	15.7	403,704	16.4	48,809	15.8	134,924	16.8	159,052	21.2
High	1,038,995	12.8	479,600	13.4	361,847	14.7	45,037	14.6	122,593	15.3	128,447	17.1
Very high	463,522	5.7	211,998	5.9	166,054	6.8	22,790	7.4	50,845	6.3	64,943	8.7

in high to very high flood risk zones, which is further compounded by the number of females, young, elderly, and illiterate people. At least 25.8% of housing units are located in these zones. As noted earlier, the study is based on 2001 data and does not take into account potential population growth in the study area, which may underestimate the actual population at risk. In addition, it was not possible to consider pluvial flooding, which is perhaps widespread in embanked areas, due to data constraints. However, the flood risk map developed here can be used as baseline information for the development of adaptation measures to mitigate future losses driven by climate change, and it can subsequently be updated with the 2011 census data upon release.

6.7 Summary

This chapter described flood vulnerability and risk assessment by using environmental and socioeconomic data. A number of physical, social, and economic factors were derived from census and spatial databases to determine flood vulnerability. To understand the potential for loss of lives and property from floods, a fine-scale spatial unit, such as the community, was considered. The chapter presented a method for flood vulnerability assessment in a data-poor country by harnessing spatial MCE techniques in terms of the AHP with WLC within a GIS. Coping capacity was conceived as part of the vulnerability and was mapped out; finally, a risk map was created by multiplying flood hazard with the composite vulnerability map.

The results demonstrated that flood vulnerability in Dhaka is the outcome of physical, social, and economic factors. More than 40% of the population lives in close proximity to active channels, which is compounded by the low elevation of the area. In line with the literature, the study confirmed that high poverty, extreme population density, poor-quality houses, and limited access to safe water and sanitation are exacerbating societal vulnerability to floods. The CVI revealed that more than 42% of the population in the study area is highly susceptible to floods for a number of reasons, including low literacy and high poverty. The intersection of variables with flood risk map suggested that more than 18% of the population is at a greater risk of floods. At least 25% of the different types of houses are exposed to a high risk of flooding. The study demonstrated the spatial variability of risk from floods, which can make a substantial contribution to the public policy arena in Dhaka or elsewhere.

References

Agarwal B (1990) Social security and the family: coping with seasonality and calamity in rural India. J Peasant Stud 17(3):341–412

Alam SMN (1990) Perceptions of flood among Bangladeshi villagers. Disasters 14(4):354–357

Alam MJB, Ali MH (2002) Concept of flood shelter to cope with flood. In: Ali MA, Seraj SM, Ahmad E (eds) Engineering concerns of flood. Bangladesh University of Engineering and Technology, Dhaka, pp 175–186

Alam MS, Chowdhury AH (1992) Perception and adjustment to flood hazards in the Meghna-Dhonagoda flood plain. Jahangirnagar Rev Part II: Soc Sci XIII & XIV:33–44
Alam S, Samsuddin D (1988) Flood in Bangladesh: causes and coexistence with flood. Bhugol Patrika (Geogr Mag) 7:5–16 (in Bengali)
Ali AMS (2004) Flood event in south-western Bangladesh: a study of its nature, causes, and human perception and adjustments to a new hazard. Nat Hazard 40(1):89–111
Anderson MB (1992) Metropolitan areas and disaster vulnerability: a consideration for developing countries. In: Kreimer A, Munasinghe M (eds) Environmental management and urban vulnerability. The World Bank, Washington, DC, pp 77–92
Anderson MB, Woodrow PJ (1998) Rising from the ashes: development strategies in times of disaster. IT Publications, London
Ansary MA (2003) Seismic risk in urban areas of Bangladesh. Ministry of Environment and Forest, Dhaka
Apel H, Aronica GT, Kreibich H, Theiken AH (2009) Flood risks analyses – how detailed do we need to be? Nat Hazard 49(1):79–98
Azar D, Rain D (2007) Identifying population vulnerable to hydrological hazards in San Juan, Puerto Rico. GeoJournal 69(1–2):23–43
Bangladesh Bureau of Statistics (BBS) (2003) Population census – 2001. Community series – Dhaka, Narayanganj, Gazipur. Bangladesh Bureau of Statistics, Ministry of Planning, Dhaka
Bangladesh Bureau of Statistics (BBS), The World Bank, World Food Programme (WFP) (2005) Bangladesh household income and expenditure survey 2005, Dhaka
Baquee MA (1997) Coping with floods and erosion in Bangladesh Charlands. Asia Pacific J Environ Dev 4(2):38–52
Begum R (1993) Women in environmental disasters: the 1991 cyclone in Bangladesh. Focus Gend 1(1):34–39
Birkmann J (2007) Risk and vulnerability indicators at different scales: applicability, usefulness and policy implications. Environ Hazard 7(1):20–31
Bollin C, Hidajat R (2006) Community-based risk index: pilot implementation in Indonesia. In: Birkmann J (ed) Measuring vulnerability to natural hazards: towards disaster resilient society. United Nations University Press, Tokyo, pp 271–289
Botzen WJW, Aerts JCJH, van den Bergh JCJM (2009) Dependence of flood risk perceptions on socioeconomic and objective risk factors. Water Resour Res 45:W10440
Braun B, Aßheuer T (2011) Floods in megacity environments: vulnerability and coping strategies of slum dwellers in Dhaka/Bangladesh. Nat Hazard 58(2):771–787
Brouwer R, Akter S, Brander L, Haque E (2007) Socioeconomic vulnerability and adaptation to environmental risk: a case study of climate change and flooding in Bangladesh. Risk Anal 27(2):313–326
Burton I, Kates R, White GF (1993) The environment as hazard, 2nd edn. Guilford Press, New York
Cannon T (2002) Gender and climate hazards in Bangladesh. Gend Dev 10(2):45–50
Chakraborty J, Tobin GA, Montz BE (2005) Population evacuation: assessing spatial variability in geophysical and social vulnerability to natural hazards. Nat Hazard Rev 6(1):23–33
Choudhury NY, Paul A, Paul BK (2004) Impact of costal embankment on the flash flood in Bangladesh: a case study. Appl Geogr 24(3):241–258
Chowdhury AMR, Bhuyia AU, Choudhury AY, Sen R (1993) The Bangladesh cyclone of 1991: why so many people died. Disasters 17(4):291–304
Cook BR (2010) Flood knowledge and management in Bangladesh: increasing diversity, complexity and uncertainty. Geogr Compass 4(7):750–767
Cova TJ, Church L (1997) Modeling community vulnerability evacuation using GIS. Int J Geogr Inf Sci 11(8):763–784
Cutter SL, Mitchell JT, Scott MS (2000) Revealing the vulnerability of people and places: a case study of Georgetown County, South Carolina. Ann Assoc Am Geogr 90(4):713–737
Cutter SL, Boruff BJ, Shirley WL (2003) Social vulnerability to environmental hazards. Soc Sci Q 84(2):242–261

Cutter SL, Emrich CT, Webb JJ, Morath D (2009) Social vulnerability to climate variability hazards: a review of literature. Final report to Oxfam America. Available at http://adapt.oxfamamerica.org/resources/Literature_Review.pdf

D'Oyley V, Blunt A, Barnhardt R (1994) Education and development: lessons from the Third World. Detselig Enterprises, Calgary

Das SC (1988) Floods in Bangladesh: socio economic impact. Bhugol Patrika (Geogr Mag) 7:17–23 (in Bengali)

Dasgupta A (2007) Floods and poverty traps: evidence from Bangladesh. Econ Pol Wkly 42(30):3166–3171

Davis I (1987) Safe shelter within unsafe cities. Open House Int 12(3):5–15

del Ninno C, Dorosh PA, Smith LC (2003) Public policy, markets and household coping strategies in Bangladesh: avoiding a food security crisis following the 1998 floods. World Dev 31(7): 1221–1238

DeSteiguer JE, Duberstein J, Lopes V (2003) The analytic hierarchy process as a means for integrated watershed management. In: Proceedings of the first interagency conference on research in the watersheds, US Department of Agriculture, pp 736–740

Dewan AM (2006) Flood risk perception in Greater Dhaka: an empirical study. Geogr Environ J 5(1):160–171 (in Bengali)

Dewan AM, Corner RJ (2012) The impact of land use and land cover changes on land surface temperature in a rapidly urbanizing megacity. In: IGARSS 2012, Munich, Germany

Dewan AM, Asaduzzaman S, Mostafa G (1999) Floods in Bangladesh and co-existence of people: experience from two severely flooded areas in 1998 and 1988. Deve Dialogue (Unnayan Bitarka) 18(1):105–117 (in Bengali)

Dewan AM, Kabir MH, Nahar K, Rahman MZ (2012) Urbanization and environmental degradation in Dhaka metropolitan area of Bangladesh. Int J Environ Sustain Dev 11(2):118–146

Dodson B (1996) Vulnerability to flooding in rural Bangladesh: a socioeconomic appraisal. In: Singh RB (ed) Disasters, environment and development (Proceedings of International Geographical Union seminar, New Delhi, 9–12 December 1994). Oxford/IBH Publication, New Delhi/Calcutta, pp 183–195

Economic and Social Commission for Asia and the Pacific (ESCAP) (1991) Manual and guidelines for comprehensive flood loss prevention and management. United Nations Development Programme (UNDP), New York

Etkin D (1999) Risk transference and related trends: driving forces towards more mega-disasters. Environ Hazard 1(2):69–75

Fekete A, Damm M, Birkmann J (2010) Scales as a challenge for vulnerability assessment. Nat Hazard 55(3):729–747

Few R (2003) Flooding, vulnerability and coping strategies: local responses to a global threat. Prog Dev 3(1):43–58

Flax LK, Jackson RW, Stein DN (2002) Community vulnerability assessment tool methodology. Nat Hazard Rev 3(4):163–176

Hasan M, Haque SM, Saroar MM (1999) Responses and perceptions of people while coping with floods and riverbank erosion in Bangladesh. Khulna Univ Stud 1(1):339–347

Hewitt K (1997) Regions of risk: a geographical introduction to disasters. Longman, Essex

Holt JB, Lo CP, Hodler TW (2004) Dasymetric estimation of population density and areal interpolation of census data. Cartogr Geogr Inf Sci 31(2):103–121

Hossain S (2008) Rapid urban growth and poverty in Dhaka city. Bangladesh e-J Soc 5:1–24

Hutton D, Haque CE (2004) Human vulnerability, dislocation and resettlement: adaptation process of river bank erosion-induced displaces in Bangladesh. Disasters 28(1):41–62

Ikeda K (1995) Gender differences in human loss and vulnerability natural disasters: a case study from Bangladesh. Indian J Gend Stud 2(2):171–193

Islam MA (1974) Tropical cyclones: coastal Bangladesh. In: White GF (ed) Natural hazards: local, national, global. Oxford University Press, New York, pp 19–24

Islam MZ (1990) Failure of flood embankments: case studies of some selected projects in Bangladesh. Final report, R02/91. Institute of Flood Control and Drainage Research (IFCDR), BUET, Dhaka
Islam N, Sultana NY (2006) The status of women in Bangladesh: is the situation really encouraging. Res J Soc Sci 1:56–65
Islam K, Yusuf HKM (2001) Impact of the 1998 flood on surface and drinking water quality of Dhaka City in Bangladesh. J Asiatic Soc Bangladesh (Science) 27(1):1–8
Kaniasty K, Norris FH (1999) The experience of disaster: individuals and communities sharing trauma. In: Gist R, Lubin B (eds) Response to disaster. Taylor and Francis, Philadelphia, pp 25–62
Khan MMI (1991) The impact of local elites on disaster preparedness planning: the location of flood shelters in northern Bangladesh. Disasters 15(4):340–354
Khan MSA, Rahman MR, Chowdhury JU (2005) Stormwater quality in selected catchments in Dhaka City. Bangladesh J Water Resour Res 20:129–138
Khandker SR (2007) Coping with flood: role of institutions in Bangladesh. Agric Econ 36(2):169–180
Khatun H (2005) Flood warning and preparedness: women managing floods in Bangladesh. Orient Geogr 49(1):61–73
Khondker HH (1996) Women and floods in Bangladesh. Int J Mass Emerg Disast 14(3): 281–292
Kondo H, Seo N, Yasuda T, Hashizume M, Koido Y, Yamamoto Y (2002) Post-flood epidemics of infectious disease in Mozambique. Prehosp Disaster Med 17(3):126–133
Kubal C, Haase D, Meyer V, Scheuer S (2009) Integrated urban flood risk assessment – adapting a multi-criteria approach to a city. Nat Hazard Earth Syst Sci 9(6):1881–1895
Liverman DM (1990) Drought in Mexico: climate, agriculture, technology and land tenure in Sonora and Puebla. Ann Assoc Am Geogr 80(1):49–72
Lo CP, Faber BJ (1997) Integration of Landsat Thematic Mapper and census data for quality of life assessment. Remote Sens Environ 62(2):143–157
Malczewski J (1996) A GIS-based approach to multiple criteria group decision making. Int J Geogr Inf Syst 10(8):955–971
Malczewski J (1999) GIS and multiple-criteria decision analysis. Wiley, New York
Malczewski J (2004) GIS-based land-use suitability analysis: a critical overview. Prog Plan 62(1):3–65
Malczewski J (2006) GIS-based multi-criteria decision analysis: a survey of the literature. Int J Geogr Inf Sci 20(7):703–726
Maniruzzaman KM, Alam BM (2002) A study on the disaster response for shelters during the 1998 flood in Dhaka city. In: Ali MA, Seraj SM, Ahmad E (eds) Engineering concerns of flood. Bangladesh University of Engineering and Technology, Dhaka, pp 187–200
Maskrey A (1989). Disaster mitigation: a community based approach. Development guidelines no. 3. Oxfam, Oxford
Masood M, Takeuchi K (2012) Assessment of flood hazard, vulnerability and risk of mid-eastern Dhaka using DEM and 1D hydrodynamic model. Nat Hazard. doi:10.1007/s11069-011-0060-x
Meyer V, Scheuer S, Haase D (2009) A multicriteria approach for flood risk mapping exemplified at the Mulde river, Germany. Nat Hazard 48(1):17–39
Milly PCD, Betancourt J, Falkenmark M, Hirsch RM, Kundzewicz ZW, Lettenmaier DP, Stouffer RJ (2008) Stationarity is dead: whither water management? Science 319:573–574
Morrow BH (1999) Identifying and mapping community vulnerability. Disasters 23(1):1–18
Ngo EB (2001) When disasters and age collide: reviewing vulnerability of the elderly. Nat Hazard Rev 2(2):80–89
O'Brien P, Mileti D (1992) Citizen participation in emergency response following the Loma Prieta earthquake. Int J Mass Emerg Disast 10:71–89
Paul BK (1984) Perception and agricultural adjustments to floods in Jamuna floodplain, Bangladesh. Hum Ecol 12(1):3–19

Paul BK (1995) Farmers' responses to the Flood Action Plan (FAP) of Bangladesh – an empirical study. World Dev 23(2):299–309
Paul BK (1997) Flood research in Bangladesh in retrospect and prospect: a review. Geoforum 28:121–131
Paul BK (2003) Relief assistance to 1998 flood victims: a comparison of the performance of the government and NGOs. Geogr J 169(1):75–89
Paul BK (2010) Human injuries caused by Bangladesh's cyclone SIDR: an empirical study. Nat Hazard 54(2):483–495
Paul BK, Bhuiyan RH (2010) Urban earthquake hazard: perceived seismic risk and preparedness in Dhaka City, Bangladesh. Disasters 34(2):337–359
Paul SK, Routray JK (2010) Flood proneness and coping strategies: the experiences of two villages in Bangladesh. Disasters 34(2):489–508
Paul SK, Routray JK (2011) Household response to cyclone and induce surge in coastal Bangladesh: coping strategies and explanatory variables. Nat Hazard 57(2):477–499
Pelling M (1997) What determines vulnerability to floods: a case study in Georgetown, Guyana. Environ Urban 9(1):203–226
Quraishi SB, Choudhury MN, Khan MMK, Mia MY, Akhter S (2005) Studies on water quality parameters in flood water. J Bangladesh Acad Sci 29(2):195–200
Rahman A (1996) Peoples' perception and response to flooding: the Bangladesh experience. J Conting Crisis Manage 4(4):198–207
Rahman MM, Bux MK (1995) Post disaster situation of water supply and sanitation. J Civil Eng Div IEB Bangladesh CE23(1):91–102
Rahman MM, Rahman MM, Hossain MD (2002) Assessment of water quality in flood affected areas of Dhaka City. In: Ali MA, Seraj SM, Ahmad S (eds) Engineering concerns of flood. Bangladesh University of Engineering and Technology (BUET), Dhaka, pp 35–47
Rahman MR, Azad AK, Nahar SB, Ali A, Hossain M, Jahan Y, Ara R, Hoque MM (2004) The impact of floods on health and hygiene in rural and urban areas and mitigation options. In: Siddiqui KU, Hossain ANH (eds) Options for flood risk and damage reduction in Bangladesh. The University Press Limited, Dhaka, pp 383–398
Ramanathan R (2001) A note on the use of the analytic hierarchy process for environmental impact assessment. J Environ Impact Assess 63(1):27–35
Rashid SF (2000) The urban poor in Dhaka city: their struggles and coping strategies during the floods of 1998. Disasters 24(3):240–253
Rasid H, Mallik A (1995) Flood adaptations in Bangladesh: is the compartmentalization scheme compatible with indigenous adjustments of rice cropping to flood regimes? Appl Geogr 15(1):3–17
Rasid H, Haider W, Hunt L (2000) Post flood assessment of emergency evacuation policies in the Red river delta, Southern Manitoba. Can Geogr 44(4):369–386
Ray-Bennett NS (2009) Coping with multiples disasters and diminishing livelihood resources: caste, class and gender perspectives: the case from Orissa, India. Reg Dev Dialogue 30(1):108–120
Rayhan IM (2008) Assessing household vulnerability and coping strategies to floods: a comparative study of flooded and non-flooded areas in Bangladesh, 2005. Cuvillier Verlag, Gottingen
Saaty TL (1977) A scaling method for priorities in hierarchical structures. J Math Psychol 15(3):234–281
Saaty TL (1980) The analytical hierarchy process. McGraw-Hill, New York
Sanyal J, Lu XX (2006) GIS based flood hazard mapping in Gangetic West Bengal. Singap J Trop Geogr 27(2):207–220
Sanyal J, Lu XX (2009) Ideal location for flood shelter: a geographic information system approach. J Flood Risk Assess 2(4):262–271
Satterthwaite D (2010) Avoiding the urbanization of disasters. In: World disasters report – focus on urban risk. International Federation of Red Cross and Red Crescent Societies, Geneva, pp 11–29
Shaw R (1989) Living with floods in Bangladesh. Anthropol Today 5(1):10–13
Siddique AK, Islam Q, Akram K, Mazumder Y, Mitra A, Eusof A (1989) Cholera epidemic and natural disaster; where is the link. Trop Geogr Med 41(4):377–382

Sohel KMA, Chowdhury MAI, Ahmed MF (2003) Surface water quality in and around Dhaka City. J Water Supply Res Technol–AQUA 52(2):141–153
Sultana F (2010) Living in hazardous waterscapes: gendered vulnerabilities and experiences of floods and disasters. Environ Hazard 9(1):43–53
Sultana P, Thompson P (2008) Gender and local floodplain management institutions: a case study from Bangladesh. J Int Dev 20(1):53–68
Takeuchi K (2001) Increasing vulnerability to extreme floods and societal needs of hydrological forecasting. Hydrol Sci J 46(6):869–881
Tapsell SM, Penning-Rowsell C, Tunstall SM, Wilson TL (2002) Vulnerability to flooding: health and social dimensions. Philos Trans R Soc Lond A 360(1796):1511–1525
Tingsanchali T, Karim MF (2005) Flood hazard and risk analysis in the southwest region of Bangladesh. Hydrol Process 19(10):2055–2069
Tobin GA, Montz BE (1997) Natural hazards: explanation and integration. Guilford Press, New York
Tran P, Shaw R, Chantry G, Norton J (2009) GIS and local knowledge in disaster management: a case study of flood risk mapping in Vietnam. Disasters 33(1):152–169
United Nations (UN) (2003) Guidelines for reducing flood losses. Available at http://www.un.org/esa/sustdev/publications/flood_guidelines.pdf. Accessed 12 Jan, 2010
United Nations Development Programme (UNDP) (1992) An overview of disaster management, 2nd edn. UNDP, Washington, DC
UN/ISDR (International Strategy for Disaster Reduction) (2004) Living with risk: a global review of disaster reduction initiatives. United Nations Publications, Geneva
Wang Y, Li Z, Tang Z, Zeng G (2011) A GIS-based spatial multi-criteria approach for flood risk assessment in Dongting lake region, Hunan, Central China. Water Resour Manage 25(13):3465–3484
Wisner B (2003) Disaster risk reduction in megacities: making the most of human and social capital. In: Kreimer A, Arnold M, Carlin A (eds) Building safer cities: the futures of disaster risk. The World Bank, Washington, DC, pp 181–196
Wisner B, Blaikie P, Cannon T, Davis I (2004) At risk: natural hazard, people's vulnerability and disasters, 2nd edn. Routledge, Abingdon
World Bank (2011) Urban risk assessments: an approach for understanding disaster and climate risk in cities. Urban Development and Local Government Unit, Washington, DC
Yoon DK (2012) Assessment of social vulnerability to natural disasters: a comparative study. Nat Hazard 63(2):823–843
Yoshino F, Yoshikawa K (1985) A study on flood risk mapping. In: Proceedings of the international symposium on erosion, debris flow and disaster prevention, Tsukuba, Japan, pp 499–504
Younus MA, Bedford RD, Morad M (2005) Not so high and dry: patterns of 'autonomous adjustment' to major flooding events in Bangladesh. Geography 90(2):112–120
Zaman MQ (1993) Rivers of life: living with floods in Bangladesh. Asian Surv 33(10):985–996

Chapter 7
Conclusions and Recommendations

7.1 Introduction

Flood has been singled out because it continues to be a matter of concern in many parts of the world, as suggested by various climate predictive models. Even though both rural and urban communities are equally prone to flood hazards, literature suggest that the effect of floods is enormous in urban areas.

The vulnerability of megacities to floods is accelerating, particularly in developing countries, primarily due to the large concentration of human populations and dense infrastructures. Further, unplanned development, rampant poverty, weak environmental legislation, inability to cope with environmental hazards, locational exposure, political structure, and climate warming contribute to the increasing flood risk. Although various adaptation measures have been taken to ameliorate flood vulnerability in various megacities around the world, existing engineering works may not be effective in the event of intensified precipitation resulting from climate warming, simply because these flood-control works are based on stationarity (Milly et al. 2008).

Accurate information on past flood events is particularly important in order to predict the likely pattern of risks from environmental hazards. As indicated earlier, this information is lacking in many megacities in developing countries, including Dhaka, the capital of Bangladesh. This study developed spatial information on hazard, risk, and vulnerability in relation to flooding. Flood events that occurred between 1988 and 2009 were taken into account, and spatial multi-criteria evaluation techniques within a geospatial approach were considered in order to understand flood risk at the lowest level of census geography (i.e., at the community scale).

This chapter presents major findings of this work and recommendations for further research. Section 7.2 discusses the major findings, while the limitations of this work are elucidated in Sect. 7.3, and the potential applications of this study, along with research demand, are discussed in Sect. 7.4.

A.M. Dewan, *Floods in a Megacity: Geospatial Techniques in Assessing Hazards, Risk and Vulnerability*, Springer Geography, DOI 10.1007/978-94-007-5875-9_7,

7.2 Major Findings

7.2.1 Chapter 1

Chapter 1 presented an overview of floods in the global and South Asian context. Floods in Bangladesh and their management aspects were examined. In addition, vulnerability of megacities to natural hazards was investigated with special reference to urban flooding.

Between 1950 and 2011, 7,899 hydrometeorological events were reported around the world, of which 3,954 were floods of various types. Among the flood events, 52.2% occurred during 2000–2011. The same figure represents only 2% of the total flood events between 1950 and 1959. The analysis also suggested that floods killed 2.3 million people around the world and affected 3.5 billion people between 1950 and 2011. Among the total deaths attributed to floods between 1950 and 2011, 96% occurred in Asia, 2.6% in the Americas, 0.9% in Africa, 0.4% in Europe, and 0.02% in Oceania. During 2000–2011, 68.8% of flood deaths occurred on the Asian continent, followed by the Americas (15.7%).

Although fatalities related to floods have significantly declined, economic losses have increased concurrently. For instance, floods that occurred between 2000 and 2011 resulted in an estimated loss of more than US$285 billion. Conversely, flood-related losses totaled $1.8 billion during 1950–1959. However, economic damage was disproportionate among continents. Asia experienced the highest losses from floods between 1950 and 2011, amounting to more than 60% of the global damage, followed by Europe (19%), the Americas (16.8%), Oceania (2.5%), and Africa (1.2%).

South Asia, one of the most impoverished regions of the world, suffers from floods every year, with an average yearly incidence of nine. A number of internal and external factors exacerbate the flood problems in South Asia. Floods in the region killed 0.14 million people and affected 1.2 billion people between 1950 and 2011. Consistent with global estimates, economic losses due to floods are increasing. Losses from floods in the region totaled $65.3 billion, of which India had the highest economic damage (54.9%), followed by Pakistan (22.9%) and Bangladesh (18.5%). The spatial distribution of floods showed that India ranks on top and Bangladesh ranks on the second.

Located in the downstream part of the Ganges–Brahmaputra–Meghna basin, Bangladesh is one of the world's most exposed countries to the risk of floods. While 20–30% of the land is subject to normal inundation every year, this figure could rise to more than 60% in the event of a catastrophe, resulting in massive economic losses and considerable human suffering. It is estimated that, every year, Bangladesh loses US$175 million from flood-related disasters, which could be in excess of US$2 billion during calamitous floods. A mixture of physical and anthropogenic factors is accountable for exacerbating flooding in Bangladesh. To save lives and property from floods, a number of flood-control and drainage structures have been built since 1960s; however, flood damage continues to rise. For instance, the 2004 flood, which is believed to be a moderate event, caused economic losses of US$2.3 billion, which was higher than the 1998 floods—the worst on record. Conversely, environmental degradation in relation to flood-control structures has become severe throughout the country. In addition, flood hazard areas have increased markedly. The problem of flooding is likely to be exacerbated under warmer climate, as precipitation is predicted to increase.

The number of megacities around the world is increasing, and the majority will be located in developing countries. While the world had two megacities in 1970, the figure is currently 23. Though megacities act as engines of economic and social improvement to countries, they have become the most risky areas and hot spots of environmental hazards, particularly in developing countries. A number of factors, such as high population growth, physical exposure, inappropriate urban development, increasing poverty, and poor-quality houses are being contributed to the increased risk of megacities to environmental hazards in low-income countries.

Urban flooding has become a common problem in megacities around the world. Due to high population density and increasing concentration of economic activities, particularly in developing countries, megacities in these nations are currently facing tremendous challenges to ameliorate flood-related losses and vulnerability. Estimates show that at least 233 cities around the world are exposed to high risk of flood, potentially affecting 663 million people. As impervious surfaces such as roofs, roads, and pavements increase with the expansion of built-up surfaces, flooding in urban areas is intensified. The level of urbanization is expected to increase in Asian megacities, which is likely to intensify flood losses. There are four types of urban flooding: fluvial, pluvial, coastal, and flash floods. Depending on a megacity's location and drainage infrastructures, floods can devastate its population and property.

7.2.2 Chapter 2

Assessment of vulnerability to natural hazards has received ample attention in recent times in order to understand human susceptibility to environmental risk. Consequently, a number of models have been developed, and they were reviewed. Based on these models, this study developed a framework for vulnerability assessment. Although a variety of definitions exist regarding the notion of vulnerability, this study conceived that the vulnerability to natural hazards is the combination of physical and socioeconomic factors and that hazard is a pre-existing condition, while risk is viewed as the product of hazard and vulnerability. To determine vulnerability of an area (e.g., city/megacity), the current state of the communities coping capacity should be analyzed, and it should not be viewed as a separate element. Rather, a community's ability to withstand or recover from hazards should be included in the evaluation of vulnerability. Thus, total vulnerability can be measured and incorporated into risk assessment.

Geospatial techniques are increasingly being used to determine vulnerability in various settings, including city/megacity, to natural hazards (e.g., floods). They can be categorized into three major domains: flood mapping and monitoring, damage assessment, and flood risk assessment, which includes hazard and vulnerability estimations. Although these techniques are able to contribute substantially to any of the four phases of disaster management, a variety of approaches are being used to evaluate vulnerability and risk of urban areas (e.g., city/megacity). Flood maps obtained from spaceborne and airborne remote sensing (RS) are of particular importance to the accurate delineation of flood-prone areas, which can further be integrated with urban land-use data to estimate flood damage. Using the spatial analytical capabilities of a GIS, several spatial coexistence models have been developed for the assessment of flood-related risk and vulnerability in diverse environments. The basic premise of these studies is to systematically identify vulnerable people and property for the reduction of risk caused by floods.

7.2.3 Chapter 3

The megacity of Dhaka, which is the capital of Bangladesh, first attracted attention when it became the provincial capital in 1905. After the country gained independence in 1971, it became the major focus of administrative, social, educational, and cultural activities. Consequently, a tremendous growth in population was observed. Currently, the population of Dhaka exceeds 14 million, and it has an average annual growth rate of 4.08%. Understandably, these people exert tremendous pressure on land for housing and other urban services, which in turn has profound environmental implications, including floods.

Historically, the city was endowed with natural drainage infrastructures such as natural waterways, canals, and intricate river systems. With rapid urban expansion driven by population growth, many of these natural waterways have been filled up. Although storm water facilities are in place to ameliorate flood damage, they are inadequate in conveying massive volumes of monsoonal flow; thereby, the vulnerability of floods is increasing.

There have been a number of disastrous floods in Dhaka since 1950. These floods were mainly caused by monsoonal downpours and the spilling of surrounding river waters. However, pluvial flooding has now become a grave concern to the city's authority and its inhabitants, as this type of flood severely affects the regular functioning of urban systems during the wet season. Although the floods that occurred in the 1970s were not significantly different from those that occurred in the recent past, from a hydrological point of view, increasing flood damage in Dhaka is primarily attributed to the growing occupancy of flood-prone areas and the development of urban infrastructure without taking into account the effects of land-use change on hydrology.

The study demonstrated that at least 10 major factors are accountable for Dhaka's increasing flood vulnerability. They include: (1) low elevation, (2) unplanned urbanization, (3) dilapidated drainage systems, (4) poor performance of flood-control works, (5) high population density and extreme poverty, (6) flood preparedness, (7) sociopolitical structure, (8) governance quality, (9) effects of climate change, and

(10) the absence of risk communication tools. As urban planning and related developments are based on traditional policies, many of the proposed adaptation measures will not work because the city already has an *adaptation deficit*, as non-climatic factors such as ill-structured urban growth are consistently compelling people to settle in dangerous locations, thereby increasing exposure to flood hazards.

7.2.4 Chapter 4

Using spaceborne RS data, flood mapping from 1988 to 2009 was performed. In addition, floodwater depths were calculated by making use of a DEM with the HWL of five major rivers flowing to the study area. Flood damage assessment was conducted, and synthetic depth–damage curves were derived for housing and road networks. Finally, flood maps were subject to accuracy assessment.

The spatial distribution of floods demonstrated that the Dhaka megacity was heavily flooded during 1988 and 1998. At least 43% of the area was flooded in 1988. The flood that occurred in 1998 was the worst on record, inundating 42.5% of the study area despite massive flood-control works in the 1990s. The moderate flood events of 2004 and 2007 inundated 35 and 37% of the study area, respectively. The study also indicated that at least 25% of Dhaka is prone to inundation in a normal period. The overall accuracy of the flood maps ranged between 68 and 75%.

Flood damage estimation in terms of repair costs for housing showed that *katcha* and *semi-pucca* houses suffered the most damage during the 2004 and 2007 floods. As these housing units usually shelter low-income and lower-middle-income groups in the study area, the potential damage from future events remains high. Conversely, road damage during the 1998 and 2007 floods was almost identical, despite the fact that the 2007 flood was a moderate event compared with the 1998 flood. A depth–damage curve for housing and roads was developed, which indicated that flood loss is rising despite major flood-control initiatives, signifying that existing policies for flood hazard mitigation need to be reevaluated, as precipitation is expected to be intensified in the coming years.

7.2.5 Chapter 5

The spatial extent of flood and depth maps was used to derive flood-affected frequency and floodwater depth maps. These two maps were combined, and a ranking method was used to model flood hazards in the study area. The hazard map was reclassified according to the severity of flood hazards, and the spatial distribution of hazardous areas was mapped. The hazard map revealed that peripheral zones that are being considered for urban expansion were in the high-hazard zone, while the most elevated and built-up lands were in no- to low-hazard zones.

The flood hazard map was overlaid with demographic data to determine the percentage of people exposed to different hazard categories. It was found that at least 5% of the population is exposed to a high hazard, while more than 18% is exposed to low to medium flood hazard categories. The distribution of the elderly population between medium and high hazardous areas was found to be 10% of the total population. Similarly, more than 9% of the female population was exposed to medium to high flood hazards.

The number of houses in each hazard zone showed that 35% of *katcha* houses were exposed to low- and high-hazard zones. Likewise, the total number of *semi-pucca* and *pucca* houses in the hazardous area was 20.6 and 17.9%, respectively.

7.2.6 Chapter 6

In assessing flood vulnerability, three factors were considered: physical or locational, social, and the coping capacity of communities. Using the analytic hierarchy process (AHP) with weighted linear combination (WLC) techniques, a composite flood vulnerability map was constructed. The results demonstrated that 42% of the population was located between high and very high vulnerable zones, while 45% were in the low to medium vulnerable zones. As the literacy rate is lower, this further compounded the degree of flood vulnerability in both of these zones. Further, the percentages of *katcha* houses that are usually home to poor people were higher in the very high vulnerable zone (36.4%). In contrast, the number of *semi-pucca* houses was highest in the high vulnerable zone, which is home to middle-class and lower-middle class households. More than 8.7% of the unemployed population was located between the high and very high vulnerable zones, revealing their susceptibility to recurrent flood events in the study area. At least 9% of the *pucca* houses were also in the very high vulnerable zone, suggesting that people in the study area may have few options, as the scarcity of land is compelling them to settle in highly vulnerable areas. With the forecast increase of the population in the coming years, flood vulnerability of Dhaka will increase; therefore, it is important to recognize the spatial distribution of vulnerable people and property, which would allow efficient management of future floods.

While hazards represent physical processes, flood vulnerability indicates susceptibility to damage and the risk of human lives. Multiplication of hazard and vulnerability maps led to the development of flood risk map for study area. The intersection of socioeconomic parameters with the developed flood risk zone revealed that at least 18% of the population in the study area was at a high to very high risk of floods, which is further compounded by the number of females, young, elderly, and illiterate people. More than 25% of human settlements—in terms of different housing types—were distributed between high and very high flood risk zones. As this study is based on the 2001 population census and did not attempt to account for the potential growth of population, the results depicted here may underestimate the actual number of people at risk. Therefore, the outcome of this study should be interpreted carefully. However, the results provided here can be used as baseline information for the development of pertinent countermeasures to lessen potential flood losses driven by climate warming. Emergency managers and city authorities can use this information to better prepare for flood disasters.

7.3 Limitations of the Work

The study suffered from a number of limitations. As noted earlier, reliable data is one of the most important constraints in Bangladesh, and particularly in Dhaka. As a fine-resolution DEM was lacking in the study, a moderate-resolution DEM was used, which was deemed a potential problem in analyzing flood damage and vulnerability. Another difficulty that the study encountered was reliable damage statistics. As many organizations calculate flood damage in Dhaka, it was difficult to amalgamate them to develop a spatially explicit flood damage model for different land-use categories. Although the study has attempted to derive depth–damage curves for housing and roads, they should be interpreted cautiously, as the damage statistics are based on average repair costs obtained from literature and field visits.

Some spatial data were not available in a good resolution, which restricted comprehensive analysis. For example, census tract boundaries were not up-to-date; hence, it took a significant amount of time to prepare census boundary data for integrating the population census of 2001. In addition, census data were not available in digital form, so time was needed to encode and relate with the census boundaries. One of the major problems was in estimating temporal vulnerability by using the 1991 and 2001 population census; however, because of the problem with the 1991 census data, the results could not be analyzed and reported here. However, the inclusion of time geography into the vulnerability model may also present a methodological hurdle. A concerted effort is therefore needed to make detailed geographically referenced data so that spatially explicit vulnerability models can be prepared for mitigating potential damage from future events. Finally, the model for vulnerability assessment only dealt with probable causes of vulnerability as opposed to systematic root causes.

7.4 Application of This Research and Further Research Demand

Flooding in urban areas can occur from the imposition of impermeable surfaces or direct inundation from river water together with incessant rainfall. Urban areas built in floodplains or in close proximity to rivers are particularly vulnerable to the latter type of floods. However, pluvial floods, which have become widespread in recent years, also present serious threats to people and urban infrastructures in the megacities of developing countries.

Despite the limitations listed above, the conceptual model developed in this study was operationalized by using spatial multi-criteria evaluation techniques within a GIS in a data-poor country. The study exploited spatial and census data to determine vulnerable locations, populations, and properties, which can considerably contribute to the deeper understanding of flood risk in Dhaka. This study is also expected to provide considerable management implications, including relief operations for high- and very high-risk areas during future floods. The developed hazard map can also be used to assign priority for the development of hazardous areas. Urban planners can use this information to make environmentally sound land-use decisions. The most important advantages of the developed hazard and risk maps are that they can be quickly updated if modifications occur subsequent to the original study (e.g., changes in land use). Moreover, these maps can be shared by various organizations involved in disaster management. Although the methods developed in this study are employed in Dhaka, the analytical tools and methodology can be used at any scale with similar characteristics around the world to identify the areas and people that are most vulnerable to natural hazards.

Nevertheless, the following are potential areas where geospatial techniques could play crucial role in reducing flood vulnerability and associated risk:

1. A study should be conducted on the effects of land-use change on flooding. It is assumed that flood magnitude and the extent of inundation are governed by the changes in land use in urban areas. Therefore, the underlying causes of increasing flood hazard in a megacity scale can be determined by researching the effects of land-use change on flooding. Moreover, studies on different urbanization scenarios and flooding extents need to be conducted to aid in predicting lands under water in different urbanization stages.
2. A comprehensive rainfall–runoff model with hourly or daily time stamps needs to be developed for capturing flood dynamics and seasonality. This is one of the most important areas where hydrometeorological and spatial data can be combined within a GIS to extract flood dynamics.
3. A study should be undertaken to assess the probable effects of climate change on flooding. It is anticipated that flood magnitudes similar to those that occurred in the past will be more frequent under a warming climate. Therefore, research on climate change effects may provide a deeper understanding of flooding in the study area or elsewhere.
4. Another important area to explore is the prudent rehabilitation of most vulnerable groups. As lower-income groups are the most affected by floods, it is crucial to develop models that would allow the saving of the loss of lives and property from recurrent flooding.
5. One of the secondary effects of flood is the risk of water-borne diseases. As the activity of pathogens increases during and after floods, the potential health effects of flooding can be traced by exploiting geospatial techniques. The integration of RS-derived parameters with other spatial models is known to have contributed

considerably to the understanding of disease dynamics. This area is virtually unexplored in the context of megacities, where overcrowding and flooding are increasing the risk of water-borne disease transmission.

6. Spatial modeling should be conducted to discover the optimal location for flood shelters. In addition, the delineation of evacuation routes during emergencies can be of significant help in saving lives. As megacities are places of extreme population density, analyzing the best evacuation routes with spatial analysis could assist in better visualization during and after emergencies.
7. As seen in Dhaka, unplanned urbanization with increasing poverty is forcing many people to settle in hazardous locations. As climatic shifts will bring changes in local hydrological cycles through intense rainfall, appropriate policies should be devised to limit the further encroachment of floodplains and hazardous areas. Spatial analytical tools can be used to identify suitable locations and subsequently assist in developing policies to adapt to the changing environment.
8. The study heavily relied on literature for deciding variables in the calculation of vulnerability. Further research could incorporate more variables and systematically identify the factors that contribute the most to the identification of vulnerable people and places. Attitude of various stakeholders involved in flood management may be included to develop a spatial decision support system (SDSS). As data remains a formidable issue in developing countries, high-resolution earth-observing satellites such as Worldview and RapidEye can be of significant help in deriving pertinent variables for vulnerability estimation.
9. A sensitivity analysis could be conducted by adopting different weighting schemes for various indicator variables. This may allow comparing different techniques and weighting schemes for environmental risk analysis.

Reference

Milly PCD, Betancourt J, Falkenmark M, Hirsch RM, Kundzewicz ZW, Lettenmaier DP, Stouffer RJ (2008) Stationarity is dead: whither water management? Science 319:573–574

Appendices

Appendix I: Pairwise Comparison Matrix for Physical Vulnerability Indicators

Criteria at level 3 of decision hierarchy model

	1	2	3	4	5
Land use/cover					
1. Cultivated	1	1/3	4	1/5	4
2. Rural settlement	3	1	6	1/3	6
3. Wetland and water	1/5	1/6	1	1/7	1
4. Built-up	5	3	7	1	6
5. Vegetation and bare	1/5	1/6	1	1/6	1
CR: 0.042					
Distance to active channel (km)					
1. <2	1	2	4	6	7
2. 2–3	1/2	1	3	5	7
3. 3–4	1/4	1/3	1	4	6
4. 4–5	1/6	1/5	1/4	1	3
5. >5	1/7	1/7	1/6	1/3	1
CR: 0.065					
Elevation (m)					
1. <4	1	3	5	7	9
2. 4–5.5	1/3	1	5	6	8
3. 5.5–7	1/5	1/5	1	3	5
4. 7–9	1/7	1/6	1/3	1	3
5. >9	1/9	1/8	1/5	1/3	1
CR: 0.08					
Geology					
1. Marsh	1	1/4	3	1/4	
2. Madhupur	4	1	5	2	
3. Alluvial silt	1/3	1/5	1	1/4	
4. Alluvial silt and clay	4	1/2	4	1	
CR: 0.065					

A.M. Dewan, *Floods in a Megacity: Geospatial Techniques in Assessing Hazards, Risk and Vulnerability*, Springer Geography, DOI 10.1007/978-94-007-5875-9,

Criteria at level 2 of decision hierarchy model

	1	2	3	4
1. Elevation	1	2	4	6
2. Land use	1/2	1	3	6
3. Distance to active channel	1/4	1/3	1	4
4. Geology	1/6	1/6	1/4	1
CR: 0.05				

Appendix II: Pairwise Comparison Matrix for Social Vulnerability Indicators

Criteria at level 4 of decision hierarchy model

	1	2	3	4	5
Population density (km^2)					
1. <30,000	1	1/3	1/5	1/7	1/9
2. 30,000–85,000	3	1	1/3	1/5	1/7
3. 85,000–165,000	5	3	1	1/3	1/5
4. 165,000–350,000	7	5	3	1	1/4
5. >350,000	9	7	5	4	1
CR: 0.068					
Age					
1. 0–4	1	4	8	4	
2. 5–9	1/4	1	5	3	
3. 10–14	1/8	1/5	1	1/4	
4. 60+	1/4	1/3	4	1	
CR: 0.079					
Illiteracy rate (%)					
1. <24	1	1/3	1/5	1/7	1/8
2. 24–38	3	1	1/3	1/5	1/7
3. 38–50	5	3	1	1/3	1/5
4. 50–70	7	5	3	1	1/3
5. >70	8	7	5	3	1
CR: 0.061					
Types of housing					
1. *Pucca*	1	1/5	1/8		
2. *Semi-pucca*	5	1	1/4		
3. *Katcha*	8	4	1		
CR: 0.082					
Poverty rate (%)					
1. <2.7	1	1/3	1/5	1/8	
2. 2.7–5.4	3	1	1/3	1/6	
3. 5.4–12	5	3	1	1/3	
4. >12	8	6	3	1	
CR: 0.037					

(continued)

(continued)

	1	2	3	4	5
Road density (km^2)					
1. <50	1	1/3	1/4	1/6	1/7
2. 50–100	3	1	1/3	1/5	1/6
3. 100–200	4	3	1	1/3	1/5
4. 200–400	6	5	3	1	1/3
5. >400	7	6	5	3	1
CR: 0.068					
Sanitation					
1. Sanitized	1	1/4	1/7		
2. Other	4	1	1/4		
3. No sanitation	7	4	1		
CR: 0.067					
Water sources					
1. Tap	1	1/3	1/5	1/6	1/7
2. Tube well	3	1	1/2	1/5	1/6
3. Well	5	2	1	1/5	1/6
4. Pond	6	5	5	1	1
5. Others	7	6	6	1	1
CR: 0.066					

Criteria at level 3 of decision hierarchy model

	1	2	3	4
Demographic				
1. Population density	1	3	5	
2. Age	1/3	1	4	
3. Gender	1/5	1/4	1	
CR: 0.075				
Socioeconomic				
Illiteracy	1	1/4	1/3	
Poverty	4	1	2	
Housing type	3	1/2	1	
CR: 0.016				
Infrastructure and lifelines				
Road density	1	1/4	1/3	
Water supply	4	1	1	
Sanitary	3	1	1	
CR: 0.008				

Criteria at level 2 of decision hierarchy model

	Demographics	Socioeconomic	Infra and lifelines
Demographic	1	2	4
Socioeconomic	1/2	1	3
Infrastructure and lifelines	1/4	1/3	1
CR: 0.01			

Appendix III: Pairwise Comparison Matrix for Coping Capacity Indicators

Criteria at level 3 of decision hierarchy model

	1	2	3	4	5
Literacy rate (%)					
1. <38	1	1/3	1/5	1/6	1/7
2. 38–52.5	3	1	1/3	1/4	1/7
3. 52.5–65	5	3	1	1/3	1/5
4. 65–76.5	6	4	3	1	1/3
5. >76.5	7	7	5	3	1
CR: 0.068					
Mean housing price ('000 BDT)					
1. <26.8	1	1/3	1/5	1/7	1/9
2. 26.8–200.09	3	1	1/3	1/6	1/8
3. 200.09–1017.3	5	3	1	1/4	1/6
4. 1017.3–3,737	7	6	4	1	1/3
5. >3,737	9	8	6	3	1
CR: 0.073					
Number of hospitals (per 1,000 people)					
1. <3	1	1/2	1/5	1/7	1/8
2. 3–10	2	1	1/4	1/6	1/8
3. 10–30	5	4	1	1/4	1/5
4. 30–100	7	6	4	1	1/3
5. >100	8	8	5	3	1
CR: 0.07					
Number of flood shelters (per 1,000 people)					
1. <2	1	1/2	1/5	1/7	1/8
2. 2–5.7	2	1	1/3	1/6	1/8
3. 5.7–13	5	3	1	1/3	1/5
4. 13–25	7	6	3	1	1/3
5. >25	8	8	5	3	1
CR: 0.047					
Flood awareness					
1. <2	1	1/2	1/5	1/7	
2. 3–4	2	1	1/3	1/5	
3. 5–6	5	3	1	1/3	
4. >7	7	5	3	1	
CR: 0.025					

Criteria at level 2 of decision hierarchy model

	1	2	3	4	5
1. Literacy rate	1	2	1/3	1/4	1/6
2. Housing price	1/2	1	1/5	1/7	1/9
3. Number of hospitals	3	5	1	1/2	1/4
4. Number of shelters	4	7	2	1	1/2
5. Flood awareness	6	9	4	2	1
CR: 0.016					

Index

A
Accuracy assessment
  error matrix, 124
  kappa coefficient, 53, 114
  mapping, 177
  overall, 122, 181
  producers, 119
  users, 119
ActionAid, 18
Adam, S., 53, 55, 123
Adana-Ceyhan, Turkey, 13
Adaptation deficit, non-climatic factor, 93, 177
Afghanistan, 5
Africa, 3, 4, 174
Akyurek, Z., 59, 60
Ali, A., 52
Americas, 3, 4, 174
Anthropogenic factors, endogenous drivers, 46
Australia, 15, 55

B
Balu, 76, 84, 121
Bangladesh, 4, 18, 75, 104, 133, 141, 173
Barton, I.J., 52
Barua, S., 8, 12, 86, 92, 122
Bathols, J.M., 52
Baumann, P.R., 51
Beijing, 11, 12, 77
Bhutan, 5
Blaikie, P., 38, 40, 44
Blasco, F., 52
Bohle, H.G., 37, 38
Brahmaputra, 9
Breaching, 9, 18, 86
Brivio, P.A., 53, 54, 56
Buenos Aires, 12, 16
Buriganga, 76, 85, 90

C
Census tracts
  districts, 105
  geography, 105, 149, 179
Chatterjee, M., 13
Chile, 13, 16
China, 16, 52, 56, 61
Cities, 6, 10–15, 18, 19, 77, 87, 175
Climate change
  climate warming, 3, 4, 180
  climatic shift, 181
  global temperature, 3
Community
  mahalla, 106
  mauza, 106
  rural, 106
  urban, 13, 61, 106, 173
Cook, B.R., 8, 157
Coping capacity, adaptive capacity, 4, 46, 92
Corner, R.J., 82, 83, 105, 141
Cova, T.J., 48
Critical facilities
  hospitals, 46
  life lines, 46
  shelters, 155
Cutter, S.L., 10, 11, 13–15, 19, 20, 35–38, 41–43, 45, 47–49, 59, 61, 82, 149, 153, 155, 159, 162
Cyclone
  coastal flood, 6, 19
  storm surges, 6, 19

A.M. Dewan, *Floods in a Megacity: Geospatial Techniques in Assessing Hazards, Risk and Vulnerability*, Springer Geography, DOI 10.1007/978-94-007-5875-9,

D
Damage
assessment, 35, 49, 50, 55–57, 103, 104, 111–113, 176, 177
depth-damage curve, 21, 56, 58, 104, 113, 119, 120, 123, 124, 177, 179
economic loss, 3, 6, 84, 174
financial, 117, 118
flood, 7, 17, 50, 55–58, 81, 84, 93, 103, 110, 112, 113, 119–122, 174, 176, 177, 179
function, 58, 113, 123
housing, 105, 120
intangible, 112
loss function, 57
road, 57, 113, 177
tangible, 112
Dasgupta, A., 6, 9, 19, 155
Davidson, R., 43, 45
Delhi, 12, 16
Developed countries, 11, 15, 20, 50, 123, 129, 133, 141
Developing countries
developing nations, 141
low-income, 155
Dewan, A.M., 6, 7, 9, 12, 17, 53, 57, 60, 78, 79, 81–84, 87, 91, 93, 105, 108, 123, 141, 156, 157
Dhaka megacity
Capital Development Authority (RAJUK), 105
detailed area plan (DAP), 91, 105
Dhaka city corporation (DCC), 76
municipality, 76
Digital elevation model (DEM)
elevation, 50, 56
terrain, 56, 57
topographic, 50
Disaster
cycle, 35, 48, 49, 62
management, 14, 48, 49, 61, 62, 91, 156, 176, 180
mitigation, 13, 14, 38, 41, 44, 48, 61
planning, 12, 13, 17, 44, 48, 61
preparedness, 13, 14, 17, 41, 48, 49, 61, 159, 162
recovery, 42, 48, 49, 93
relief, 39, 93
Disease
communicable, 156
vector borne, 9
water borne, 13, 79, 82, 155, 156, 165, 180, 181
Distributed database, web mapping, 49
Dodson, B., 8, 157
Drainage system
artificial, 12
dilapidated, 46, 84–85, 93, 121, 124, 176
Dutt, A.K., 77

E
Eakin, H., 37, 38
Early warning system, 14, 19
Ebert, A., 53, 59, 61
El Niño, exogenous factor, 7
El Niño–Southern oscillation, 7
EM-DAT/CRED, 2, 3
Emergency
cycle, 48, 62
planning, management, 48, 61
Environment
degradation, 13, 86, 87, 91, 175
ecological, 38, 45
modification, 15, 36, 137, 184
natural, 9–11, 14, 35, 36, 46, 59, 76, 91, 180
rural, 11, 14, 78
social, 36
urban, 45
Evacuation
assistance needs, 49, 91
mapping, 19
planning, 61
Evapotranspiration, 17

F
Faber, B.J., 141
Ferreira, S., 1–3
Flood
abnormal flood, 6, 9
affected-frequency, 130–136, 177
coastal flood, 6, 19
depth, 39, 56–58, 60, 104, 105, 108, 110–114, 116, 124, 130, 131, 133, 136
duration, 84, 112
early warning system, 19
exceedance probability, 57
extent, 46, 50–51, 58, 84, 86, 107–112, 114, 119, 121, 124, 130, 136
flash flood, 6, 19, 175
flood loss/damage, 6, 9, 18, 50, 57, 93, 112, 113, 119, 120, 123, 124, 139, 162, 163, 165, 175, 177, 179
fluvial, 18, 79, 81, 85, 104, 121, 123, 124, 165
forest, 55, 123

frequency, 18, 60, 129, 130, 133, 136, 161, 162
hazard, 7, 9, 21, 46, 56–61, 82, 84, 94, 123, 124, 129, 137, 145, 146, 148, 162, 170, 173, 175, 177, 178, 180
hazard analysis, 129
hazard map, 58, 60, 129, 133, 137, 146, 162, 178
highest water level, 105
impact, 92
intensity, 6, 17, 39, 109, 163
internal, 9
inundation, 39
management, 4, 8, 91–93, 103, 123, 125, 157, 159, 181
mapping and monitoring, 50–56, 176
normal flood, 6, 9, 104, 110
overland, 19
pluvial, 19, 79, 81, 85–87, 118, 120, 121, 124, 172, 176, 179
rainfall-runoff modelling, 180
rain-induced, 6
risk management, 20, 35, 49, 50, 155
riverine flood, 7
runoff, 7, 15, 17–19, 39, 58
shelters, 61, 141, 157, 181
tidal, 124
timing, 6, 17
velocity, 6, 46, 112, 130
vulnerability, 20, 21, 35, 59–61, 75–94, 142, 155, 162, 163, 170, 172, 173, 176, 178
wetland, 56, 123
Flood Action Plan (FAP), 8, 76, 79, 85, 90, 106, 113
Flood control works
channelization, 85–87, 159
embankments, 8, 9, 85, 86, 118, 163
engineering solutions/fix, 8
river engineering, 9
sluice gates/regulators, 8, 86
structural solutions, 121
Floodplain
encroachment, 7, 12, 84, 91, 124, 181
formal and informal development, 9, 83
zoning, 19, 93
Framework
conceptual, 36, 43, 45–47, 49, 62
disaster, 40, 43–45
Hyogo, 14
Fung, T., 53, 109

G
Ganges, 6, 174
Geospatial
geographic information systems (GIS), 47–50, 56, 57, 60, 61, 143
global positioning system (GPS), 47
remote sensing (RS), 47
spatial analysis, 47
Germany, 58, 60
Gillis, A.M., 9
Global change, 17, 43
Globalization, 13, 46
Governance, 91–93, 122, 176
Greenhouse gas emission, 92
Greiving, S., 19, 37, 44, 45
Griffiths, P., 82
Gross domestic product (GDP), 11

H
Haiti, 13
Hazard
adjustment, 38, 43
assessment, 60, 61, 92, 133
environmental, 10, 11, 37, 38, 43, 46, 149, 156, 173, 175
exposure, 39, 43
identification, 44, 48
man-made, 15, 48, 49
natural, 1, 4, 9–15, 20, 35, 36, 39–41, 43–49, 57, 59, 61, 62, 79, 90, 121, 122, 124, 139, 141, 149, 152, 155, 157, 161, 174, 175, 180
technological, 36, 38
water-related, 94
Henderson, F.M., 106, 123
Hewitt, K., 10, 12, 37, 38, 41, 43–45, 87, 149
Holt, J.B., 149
Houhoulis, P.F., 52
Housing
katcha, 113, 157, 159
pucca, 157
quality, 120
semi-pucca, 157
types, 21, 106, 113, 119, 121, 131, 136, 154–155, 178, 185
Huaihe, 61
Human-environment interaction, 10, 62
Hydraulic
cycle, 17, 46, 181
leakage, 9, 86
Hyogo Framework for Action, 14

I

IDNDR, 14
Imhoff, M.L., 48, 50, 51, 53–55, 108, 123
Impervious surface
  impermeable, 17
  imperviousness, 18
India, 4, 5, 16, 60, 75, 174
Infrastructures
  buildings, 82
  elements, 46
  property, 19, 82, 175
  road network, 21, 104–106, 112, 141, 155, 177
Intergovernmental Panel on Climate Change (IPCC), 1, 3, 10
Islam, M.M., 6, 19, 52, 58, 60
Islam, M.Z., 9, 77, 90

J

Jha, A.K., 1, 10, 18, 19, 38, 93

K

Kafiluddin, A.K.M., 82
Karachi, 12, 16
Kienberger, S., 19, 47, 48, 59, 61
Kolkata, 12, 16
Kubal, C., 10, 58–60, 154

L

Lakhya, 76, 90
Landscape, 11, 20
Land use/cover, 17, 18, 46, 47, 55–59, 83, 84, 93, 104, 105, 107, 110, 112, 123, 137, 141, 145, 148, 176, 179, 180, 183, 184
Lawal, D.U., 52
LeDrew, E., 53, 109
Liu, Z., 53, 57, 108, 109
Lo, C.P., 141
Lorenz, N., 51
Los-Angeles, 12, 16
Luers, A., 37, 38

M

Macintosh, H., 57
Maldives, 5
Mallik, A., 9, 87
Mason, D.C., 54, 108, 123
Megacities
  Asian, 15, 17, 175
  developing countries, 175
Meghna, 6, 76, 174
Mexico City, 12, 16
Meyer, V., 10, 19, 20, 38, 39, 48, 59, 60, 82, 142
Michener, W.K., 52
Middle-income, 155
Mitchell, J.K., 15, 45
Mitigation, 9, 13, 14, 19, 21, 38, 41, 42, 44, 47, 48, 60, 61, 123, 139, 159, 177
Model
  decision hierarchy, 147, 148, 153, 154, 157, 159, 161, 183–186
  disaster resilience of place (DROP), 43
  disaster risk reduction, 43–44
  "hazards-of-place (HOP)", 41–43
  hydrodynamic, 58, 163
  integrated assessment multi-hazards, 44
  pressure and release (PAR), 38, 40–41
  regions of risk, 43
  spatial, 20, 180, 181
  spatial coexistence, 57, 58, 176
  UN/ISDR framework for disaster risk reduction, 44–45
Monsoon, 50, 53, 75, 76
Müller, A., 14, 15, 36, 37
Multi-criteria assessment/evaluation, 35, 60, 142, 173, 180
Multi-temporal, 21, 51, 52, 54, 57, 119, 121, 124, 130, 141, 162
Mumbai, 12, 15

N

Nagarajan, R., 51
Nepal, 5, 51
New Orleans, 15, 18
New York, 12, 16
Noble, A.G., 77
North, G.W., 50, 51

O

Oceania, 3, 4, 174
Ochi, S., 133
Odeh, D.J., 20, 37, 49
Overtopping
  breaching, 9
  failure, 9
  levee, 8

P

Pakistan, 4, 5, 16, 75–77, 174
Pandey, A.C., 59, 60

Paradigm, 20, 37, 38, 161
Parker, D.J., 36–39, 112
Pelling, M., 4, 9, 10, 12, 14, 15, 36–38, 83
Perception
  disbelief, 14
  people, 157
  risk, 13
  warning system, 14
Planning
  disaster, 61, 161
  hazard, 13
  urban, 12, 84, 93, 105, 122, 177
Political
  ecology, 10, 37
  economy, 10, 37, 38, 46, 91
  instability, 17
  socio-political, 46, 77, 91, 176
Poverty
  livelihood, 38, 116
  marginal people, 116, 139
  resource inequality, 93
  slum population, 89
Profeti, G., 57

Q
Qing, C.W., 53, 55
Queensland, 15

R
Ramanathan, R., 143
Rashed, T., 10, 38, 47–49, 59, 82
Rasid, H., 9, 87
Rejeski, D., 47, 58
Remote sensing (RS), 47, 48, 50, 52, 53, 56, 58, 59, 103, 104, 106, 114, 124, 128, 133, 136, 141, 176, 177, 180
Resilience, 14, 37, 38, 43, 61
Risk
  assessment, 14, 19–21, 40, 47–62, 92, 103, 128, 139–172, 175, 176
  communication, 20, 49, 92–93, 177
  community-based risk management, 140
  environmental, 12, 13, 20, 175, 181
  flood, 9, 36, 91, 103, 128, 136, 173
RS. *See* Remote sensing (RS)

S
Saaty, T.L., 142, 144
Sado, K., 6, 19, 52, 58, 60
Sanderson, D., 13, 77, 87, 91
San Francisco, 12, 16
Sao Paulo, 12, 16
SAR. *See* Synthetic aperture radar (SAR)
Satellites
  Advanced Very High Resolution Radiometer (AVHRR), 52
  C-band, 55, 104
  ERS, 55, 56, 123
  Geoeye, 53, 106
  Ikonos, 53
  incidence angle, 55, 104, 108, 120
  Irs-1C/1D Pan, 53
  JERS-1, 54
  Landsat, 50, 51, 53, 55, 56, 82, 107, 108
  light detection and ranging (Lidar), 50
  Liss-III, 53
  moderate resolution imaging spectroradiometer (MODIS), 53
  MOS-MESSR, 117
  optical sensor, 50, 53
  passive sensor, 55
  polarization, 54, 55, 104, 105, 123
  Quickbird, 53
  radar, 53–55, 104, 106, 108, 110, 122, 123
  Radarsat, 52, 55–57, 104, 105, 107, 123
  Satellite Pour l'Observation de la Terre (SPOT), 51–53, 56, 123
  Seasat, 54
Sebastian, S., 59, 60
Sedimentation, 17
Seoul, 12, 16
Settlements
  human, 20, 135, 136, 148, 178
  urban, 13
Shanghai, 12, 16
Sheng, Y., 52, 108
Simonovic, S.P., 58
Sinha, R., 59, 60
Skidmore, A.K., 55
Socioeconomic, 7, 10, 14, 17, 38, 39, 45, 59–61, 76–79, 82, 83, 112, 141, 144, 149, 151, 154, 156, 170, 175, 178, 185
South Asia, 4, 5, 7, 174
Spatial
  analytical techniques, 47, 176, 181
  database, 21, 105
  distribution, 4, 5, 11, 20, 62, 105, 118, 120, 159, 170, 174, 177, 178
  information, 20, 49, 60, 173
  modelling, 20, 181
  models, 20, 180
  spatial-temporal, 21
Sri Lanka, 5

Srivastava, S.K., 53, 57
Stationarity, 9
Sustainability
sustainable development, 1, 44
Synthetic aperture radar (SAR), 50, 53–57, 104, 106, 109, 123

T

Techniques
analytical hierarchy process (AHP), 60, 61, 140, 142–144, 148, 149, 155, 170, 178
deductive, 35, 47, 141
geon, 61
geospatial,
inductive, 35, 47, 61, 141
maximum value composite (MVC), 53
multi-criteria assessment/evaluation (MCA/MCE), 35, 47, 60, 142, 170, 173, 180
optimal threshold, 53, 109
pairwise comparison, 143, 144, 148, 149, 156, 157
qualitative, 57, 61, 142, 156
quantitative, 48, 57, 58, 142
ranking matrix, 133, 136
spatial multi-criteria decision, 60, 140, 142, 173, 180
weighted linear combination (WLC), 61, 140, 142, 144, 148, 162, 170, 178
weighting, 61, 162, 181
Tehran, 12, 16
Thailand
Bangkok, 15
Thematic Mapper (TM)
Landsat, 51, 55, 56, 104, 107, 108, 123
Tokyo, 12, 15, 16
Tongi Khal, 76
Townsend, P.A., 50, 53, 123
Töyora, J., 53, 55, 56, 123
Tsunami
Indian Ocean, 19
Japanese, 19
Turag, 76
Turkey
Istanbul, 12, 15, 16

U

United Nations
International Strategy for Disaster Reduction(UN/ISDR), 14, 37, 39, 44–45, 136
UN-HABITAT, 13, 89
United Nations Development Programme (UNDP), 39, 40, 45, 167
United Nations Disaster Relief Co-ordinator Office(UNDRO), 39
United Nations Environmental Programme (UNEP), 37
Urban
expansion, 46, 77, 82–84, 93, 161, 176, 177
flooding, 6, 15–18, 54, 104, 123, 174, 175, 179
physical location, 10
sprawl, 12
vulnerability, 10, 13, 14, 46, 59, 82
Urbanization
crowdedness, 90
illegal encroachment, 84, 86
piecemeal development, 12, 79, 94
population density, 57, 87, 89, 93, 165, 175, 176, 181
unplanned, 82–84, 93, 176, 181
USA, 16, 50, 61

V

van Ast, J.A., 8, 12, 86, 92, 122
Victoria, 15
Vrijling, H., 12, 79, 85–87
Vulnerability
assessment, 44, 47, 62, 141–162, 170, 175, 179
biophysical, 37, 38, 41, 42, 45, 59, 61, 82, 140, 141, 144, 145, 147
causal structure, 38
community, 18, 21, 37–39, 41–47, 49, 50, 52, 57, 58, 60–62, 122, 140, 141, 148, 149, 155, 157, 159, 161, 162, 164, 165, 167, 170, 175
composite, 60, 162, 163, 170, 178
composite vulnerability index (CVI), 59, 60, 137, 158, 168–170
degree of, 15, 79, 84, 157, 165, 167, 173
demographic, 46, 59, 60
economic, 36, 37, 44, 61, 84, 140, 154, 170, 171
human dimension, 149, 167, 170
indicators, 37, 38, 42, 47, 59, 60, 140, 141, 143, 144, 148, 149, 157, 163, 165, 167
mapping, 20, 59, 92, 140, 176
persistent, 38
physical vulnerability index (PVI), 41, 46, 59, 61, 92, 141, 144, 145, 147, 148, 162, 163, 165, 167
place-based, 142
political, 10, 37, 38, 83, 140
situational, 38

social, 41, 42, 46, 59, 61, 141, 149–156, 163–166
social vulnerability index (SVI), 61, 154, 156, 162, 167
socioeconomic, 38, 59, 60, 82, 83, 141, 144, 149, 154, 170, 175
susceptibility, 140, 141, 170, 175, 178
total/overall, 46, 140, 162, 167, 175
types, 38, 155
urban, 10, 13, 14, 46, 59, 80

W
Wang, Y., 54, 55
Water bodies
encroachment, 121
natural, 121, 231
Water development projects, 8
Water pollution, 78, 155
Watts, M.J., 37, 38
Weeks, J., 38, 47–49, 59, 61
Wisner, B., 37–41, 45, 149, 162
World Bank, 10, 14, 15, 19, 77, 89, 140

Y
Yalcin, G., 59, 60
Yamaguchi, Y., 7, 17, 53, 79, 81–84

Z
Zhou, C., 52, 53
Zhou, Y.C., 53, 55

MIX
Papier aus verantwortungsvollen Quellen
Paper from responsible sources
FSC® C105338

If you have any concerns about our products,
you can contact us on
ProductSafety@springernature.com

In case Publisher is established outside the EU,
the EU authorized representative is:
Springer Nature Customer Service Center GmbH
Europaplatz 3, 69115 Heidelberg, Germany

Printed by Libri Plureos GmbH
in Hamburg, Germany